Understanding the Analytic Hierarchy Process

Series in Operations Research

Series Editors:
Malgorzata Sterna, Marco Laumanns

About the Series

The CRC Press Series in Operations Research encompasses books that contribute to the methodology of Operations Research and applying advanced analytical methods to help make better decisions.

The scope of the series is wide, including innovative applications of Operations Research which describe novel ways to solve real-world problems, with examples drawn from industrial, computing, engineering, and business applications. The series explores the latest developments in Theory and Methodology, and presents original research results contributing to the methodology of Operations Research, and to its theoretical foundations.

Featuring a broad range of reference works, textbooks and handbooks, the books in this Series will appeal not only to researchers, practitioners and students in the mathematical community, but also to engineers, physicists, and computer scientists. The inclusion of real examples and applications is highly encouraged in all of our books.

Rational Queueing
Refael Hassin

Introduction to Theory of Optimization in Euclidean Space
Samia Challal

Handbook of The Shapley Value
Encarnación Algaba, Vito Fragnelli and Joaquín Sánchez-Soriano

Advanced Studies in Multi-Criteria Decision Making
Sarah Ben Amor, João Luís de Miranda, Emel Aktas, and Adiel Teixeira de Almeida

Handbook of Military and Defense Operations Research
Natalie Scala, and James P. Howard II

Understanding the Analytic Hierarchy Process
Konrad Kułakowski

For more information about this series please visit: https://www.routledge.com/Chapman--HallCRC-Series-in-Operations-Research/book-series/CRCOPSRES

Understanding the Analytic Hierarchy Process

Konrad Kułakowski

CRC Press
Taylor & Francis Group
Boca Raton London New York

CRC Press is an imprint of the
Taylor & Francis Group, an **informa** business
A CHAPMAN & HALL BOOK

Cover image by Łukasz Zabdyr

First edition published 2021
by CRC Press
6000 Broken Sound Parkway NW, Suite 300, Boca Raton, FL 33487-2742

and by CRC Press
2 Park Square, Milton Park, Abingdon, Oxon, OX14 4RN

Library of Congress Cataloging-in-Publication Data
Names: Kułakowski, Konrad, author.
Title: Understanding the analytic hierarchy process / Konrad Kułakowski.
Description: First edition. | Boca Raton, FL : CRC Press, 2020. | Series:
    Chapman & Hall/CRC series in operations research | Includes
    bibliographical references and index.
Identifiers: LCCN 2020021002 (print) | LCCN 2020021003 (ebook) | ISBN
    9781138032323 (hardback) | ISBN 9781315392226 (ebook)
Subjects: LCSH: Decision making--Mathematical models. | Multiple criteria
    decision making.
Classification: LCC HD30.23 .K838 2020 (print) | LCC HD30.23 (ebook) |
    DDC 003/.56--dc23
LC record available at https://lccn.loc.gov/2020021002
LC ebook record available at https://lccn.loc.gov/2020021003

ISBN: 978-1-138-03232-3 (hbk)
ISBN: 978-1-315-39222-6 (ebk)

Typeset in SFRM
by Nova Techset Private Limited, Bengaluru & Chennai, India

To my wife, Anna, and children, Justyna and Jacek.

Contents

5 Rating Scale 85

6 Inconsistency 93

Foreword

In more than fifty years, the Analytical Hierarchy Process (AHP) has gone well beyond its initial formulation and has become a body of knowledge on the hierarchical arrangement of criteria and alternatives, the use of pairwise comparisons, and the aggregation of priorities according to the hierarchy. Possibly due to the generality of the approach, the AHP framework has been a fertile ground for researchers. Just to give an idea of this wealth, one can easily count more than twenty methods to derive a weight vector from a pairwise comparison matrix, and again more than twenty indices to estimate the inconsistency of pairwise comparisons.

When Konrad, years ago, disclosed his intention to write a book on the AHP, I began to wait. I had known him for some time, so I knew that it was a matter of "when" and not a matter of "if". Of course, the more time was passing, the greater my expectations were growing. Now, after much waiting, I can say that they are completely fulfilled.

This book distills and presents the most relevant results in a unique manner since it is the book which more closely offers a formal perspective on the method, and does not avoid digressions into fields like measurement theory and graph theory. Albeit neglected in most presentations of the AHP, it is undeniable that this supplementary material helps the reader understand (as the title of the book correctly suggests) the method and not feel it slip away, like sand through the hands. I maintain that this book has two additional merits: its presentation of the AHP is *wide* and *inclusive*.

It is wide because it succeeds in presenting the AHP as an organic set of methods. If, on one hand, it is certainly impossible to present all the contributions related to the AHP, on the other hand, it is necessary to make the reader aware of their existence (and whet the appetite) by means of carefully selected references to the relevant literature.

This book is inclusive in the sense that it presents concepts whose relevance goes beyond the AHP. For example, it is nowadays recognized that the concept of pairwise comparison, here defined in the form of a pairwise comparison matrix, is pervasive in the entire field of decision analysis, and not only there. It is auspicable and reasonable to expect that this book will also serve as a bridge between different communities.

The book has two types of uses. It can be read from front matter to back matter by those with no previous knowledge of the AHP and can be seen as a "cookbook", to always keep on your shelf. Each chapter is structured in such a way that makes it a self-standing story which, with some caution on the notation, is also self-contained.

In the final analysis, I believe that, whatever type of reader you are, this book will help you understand the AHP.

Matteo Brunelli
Trento, Italy
15 March 2020

Preface

People always have to make decisions. What to eat for dinner, what book to read, whom to choose as the president of the country. When making decisions, we compare different options and try to select the best alternative: the one which is most favorable for us. When there are many alternatives differing in various aspects, the decision becomes difficult. People do not know how to deal with plenty of options to choose from, how to compare them and evaluate them at once. This observation gave rise to the multiple-criteria decision-making methods that help people to deal with the complexity of making choices. Such methods are usually a combination of methodology, good practices, and procedures. Their aim is to collect a number of individual pieces of information (observation, assessments, and judgments), transform them, and then deliver the final verdict: a clear indication to decision makers of what to choose. Very often mathematics is behind the scenes. There is a set of precise rules specifying how the collected pieces of information should be converted into a ranking indicating the best possible solution. That is the theory. In practice, when we look closely at a particular decision method, we can find spots in the seemingly flawless mathematical procedure. Sometimes the obtained result is unsatisfactory. At other times, the result is not optimal; however, it could be easily improved. The use of the method sometimes leads to surprising results, and so on. To find out why this happens, it is necessary to understand the principles according to which the method was constructed: the ideas behind a specific solution, procedure, or data transformation.

The Analytic Hierarchy Process (AHP) is one of the most well-known and perhaps the most popular multiple-criteria decision-making methods. It was proposed in 1977 by *Thomas Saaty* as the hierarchical extension of comparing alternatives in pairs. This method allows its users to calculate the ranking of all alternatives based on a series of detailed questions about comparing two selected alternatives. An advantage of this method is that we can treat it as a black box. On the one hand, one enters some numerical data; on the other hand, a ready-to-use recommendation is obtained. This apparent simplicity has certainly contributed to the market success of this approach. However, this approach requires trust in the black box. AHP has been repeatedly criticized and praised since its inception. Many authors have proposed various types of extensions, modifications, changes, and improvements. So the question is, what exactly should be inside a black box? If we are to trust the effect of its calculations, what should our trust be based on? I think that, in this case, the only source of trust in the method can be understanding its mathematical foundations. For this reason, in this book I do not avoid presenting mathematical theorems with proofs. In most cases, the mathematical proofs are quite elementary and require at most basic knowledge of algebra and discrete mathematics. Therefore, following the presented reasoning should not cause difficulties for the reader. The main emphasis in the book is on the mathematical description of AHP, pairwise comparison methods, and different ways of calculating rankings and determining the inconsistency of the input data set. However, I do not write much about the practical applications of this method. Information on this subject can be found in many publications, including books and articles written by *Thomas Saaty* himself [164, 170, 169, 171, 185, 90, 79]. There are also many publications describing specific solutions in the art market, public healthcare, sustainable business, sociology, or wind energy management [183, 182, 4, 142, 145, 202, 120]. Despite some efforts, the book is not a complete overview of the various methods and improvements associated with AHP either. Given the large number of theoretical works and the continuous development of this subject, creating a complete review would be very difficult. I hope, however, that the information contained in this book will bring the reader closer to the theoretical background discussion of the practical use of the AHP method. In the book, the reader will be able to find topics classically associated with AHP, such as inconsistency, preference aggregation methods, or incompleteness. There will also be new topics, which are mostly the result of the author's work, such as ordinal inconsistency or the HRE approach.

While writing scientific works, authors usually laboriously prepare individual components to finally present the solutions. It is different in this book. At the beginning, the AHP method is shown. Some details are omitted, such as the priority-deriving method or inconsistency. In the following chapters, these missing elements are developed and discussed. The book design was inspired by the excellent and concise *Introduction to the Analytic Hierarchy Process* by *Matteo Brunelli* [27]. Similar content layout was also adopted there.

This book may be of interest to various audiences. One of them consists of practitioners who use AHP on a daily basis and want to deepen their knowledge in this area. In particular, reading the book should make it easier for them to talk to people dealing with the theoretical aspects of the decision-making methods. I would also recommend this work, although timidly, to theoreticians. Perhaps some fragments of this book may be a source of inspiration and indicate gaps in existing theories. Finally, I would recommend this book to students. The examples in this book should allow them to easily understand many of the topics discussed here.

The ideas, issues, and solutions presented in the book are the result of over six years of work in the field of decision-making methods. During this journey, many people showed me their support, help, and interest. I would like to thank *Waldemar Koczkodaj* for introducing me to the principles of the pairwise comparison method. I would also like to express my gratitude to all those who, through discussions, meetings, or joint work, influenced the final shape of this study. In particular, they were: *Jaroslav Ramík, Jiří Mazurek, Michael Soltys, Jacek Szybowski, Antoni Ligęza, Anna Prusak, José María Moreno Jiménez, Ryszard Janicki, Sándor Bozóki, Matteo Brunelli, Michele Fedrizzi, Dawid Talaga, Ryszard Szwarc,* and many others not mentioned explicitly who took part in discussions during conferences, seminars, and lectures. The book would not have been possible without the support of the Taylor & Francis Group publishing team including: *Sarfraz Khan, Callum Fraser,* and *Mansi Kabra.* Their systematic inquiries about the manuscript upheld my faith in the sense of the intended work. Special thanks go to those who reviewed the manuscript of the book. They are *Radosław Klimek, Jiří Mazurek,* and *Matteo Brunelli.* I would also like to thank Matteo for the foreword to this book. I am grateful to them for their comments and improvements. I also thank *Ian Corkill* for his editorial and language help. Some parts of this book refer to works created under the National Science Center (NCN), Poland, grant no. 2017/25/B/HS4/01617. At this point, I would like to thank NCN once again for its support. Finally, I would like to thank my family: *Anna,* my wife, and children, *Justyna* and *Jacek.* They patiently endured my moods and the time I spent shut up in the office alone with my computer. I dedicate this book to them.

Konrad Kułakowski
Kraków, Poland
18 February 2020

Author

Konrad Kułakowski is an Associate Professor and Deputy Head of the Department of Applied Computer Science, Faculty of Electrical Engineering, Automatics, Computer Science and Biomedical Engineering at the AGH University of Science and Technology (AGH UST). He received his Ph.D. and Habilitation in computer science from AGH UST. His research interests are focused on multiple-criteria decision-making methods, including AHP, and their applications, theory and practice of the pairwise comparison method, parallel programming, and algorithms. He serves as a reviewer of many international journals in operational research and computer science. He has organized several special sessions in International Conferences on Computer Science and Operational Research. He has also served as a member of program committees at numerous international IT and OR conferences and meetings.

List of Symbols

a alternative.

A set of alternatives, where $a \in A$. Very often $A = \{a_1, \ldots, a_n\}$.

C a pairwise comparisons (PC) matrix.

C^T transposition of C.

H Hessian matrix.

$\mathcal{E}, e$ usually denotes errors. In places where there is no doubt as to the meaning, e may also denote Euler's number.

c_{ij} entry of a PC matrix at the i-th row and the j-th column.

w priority vector.

w^T transposed priority vector.

$w(a_i)$ priority of the i-th alternative.

?   undefined element of an incomplete PC matrix.

T_C directed graph of a PC matrix.

P_C undirected graph of a PC matrix.

$deg_{in}(v)$ input degree of a vertex v.

$deg_{out}(v)$ output degree of a vertex v.

$deg_{un}(v)$ undirected degree of a vertex v.

$deg(v)$ degree (total) of a vertex v.

$\oplus$ adding of two fuzzy numbers.

$\ominus$ subtraction of two fuzzy numbers.

$\odot$ multiplication of two fuzzy numbers.

$\times$ Cartesian product.

$\approx$ used instead of $=$ when compared expressions have a similar value.

$|X|$ absolute value of X. In places where there is no doubt as to the meaning, it may also denote cardinality of X.

λ eigenvalue of a matrix.

λ_{max} principal eigenvalue of a matrix, spectral radius of a matrix.

w_{max} principal eigenvector, eigenvector corresponding to λ_{max}.

CI Saaty's consistency index.

CR Saaty's consistency ratio.

ζ Kendall–Babington Smith inconsistency index.

$\binom{n}{k}$ binomial expression, where $\binom{n}{k} = \frac{n!}{(n-k)!k!}$.

det determinant of a matrix.

$\ln$ natural logarithm.

$\cos$ cosine function.

Chapter 1

AHP as a Decision-Making Method

1.1 Why we need decision-making methods

The question "why do we need decision-making methods"? is fundamental for each researcher involved in decision-making methods and systems. In the case of no response or, even worse, a negative answer, we run the risk that we must deal with unnecessary items, and as a result we will not have the funds for further research. Hence, finding a convincing answer to the posed question is absolutely essential. Luckily, the response is not so hard to find. Every day, we all face the problem of choice. Which way to get to work or to school to avoid traffic jams, where to buy food or eat lunch, etc. Sometimes the decisions we have to take are related to higher expenses. Which mobile phone, TV set, computer or car to buy, which school to send the children to, and so on. In all these cases, we have to do some work related to the formulation of our needs and constraints, review the offers, read the opinions of experts, go through reviews and rankings available on-line, ask friends and colleagues for their opinion, and so on. At the end of the day, we are usually able to choose one option from many, which we believe best suits our needs. However, do we really have a guarantee that we have made the optimal choice? Of course not. It is just better to not meditate on the decision we have just made. What would happen if we actually changed our mind? This intuitive approach to decision making sometimes works, but sometimes it does not. If we choose the wrong way to work or school, we will be late. If we buy a mobile phone which we do not like, we can always sell it, but what happens when we risk other people's money and lose it? Even worse, what if we, acting as a doctor, decide to carry out a surgical operation that is not necessary? Sometimes the wrong choice is not an option. To eliminate wrong decisions or at least to minimize the risk of making suboptimal choices, we need decision-making methods.

One of the pioneers of multiple-criteria decision-making (MCDM) methods, *Bernard Roy* indicates areas where decision-making methods can be helpful [61, p. 5]. These are:

- analysis of the situation, identification of all the parties involved, possible alternatives and consequences of decisions,

- organizing and/or structuring the decision-making process, specifying alternatives, objectives and goals,

- identifying the actors, stakeholders and experts and propose them common computational and conceptual frameworks,

- preparing recommendations based on the outcome of the computational procedures and the decision process,

- explaining the final recommendation to all the parties involved or affected by the results of the decision process.

Very often, decision problems are also complicated and making a decision requires an analysis of large amounts of data. In such a case, decision-making methods or, in general, MCDM methods can come to the rescue. Usually, these are methods based on machine learning, pattern recognition or big data analysis. One of the most important advantages of MCDM methods is formalization of the decision-making process, and therefore making them "ceteris paribus" repeatable and in some sense predictable. This is crucial to justifying the achieved recommendations. Following the Latin rule *modus ponendo ponens*, since we accept that the decision method / procedure is correct and right, providing that the input data are correct, then we must also acknowledge that the results are correct and sound. Of course, we have to remember that there is no perfect MCDM method and we may always face unusual problems that either are not fully covered by our model or, for some reason in this particular case, the used methods are error-prone. Thus, using even the best MCDM method does not exempt us from the duty of using common sense fed by our professional expertise. The MCDM methods and the whole mathematic theory behind them should help us to obtain the correct decisions, but it cannot take decisions instead of us. Probably for these reasons, professionals involved in MCDM sometimes prefer to talk about decision aiding and recommendations rather than decision making and decisions. Experts and MCDM professionals may help to prepare recommendations but, at the end of the process, the outcome has to be filtered through a sieve of common sense, and then the final decision should be rendered.

In this book, we will focus mainly on the theoretical and computational aspects of pairwise comparisons (PC) and *Saaty's Analytic Hierarchy Process* (AHP). Therefore, we will look at PC and AHP as experts or MCDM professionals and scientists, rather than as the actual decision makers. One may ask whether we can be a good car driver without any knowledge of the construction and operation of the car. Of course we can. But what if the car stops in the middle of the desert and does not want to move? In this situation, knowing how the engine works in the car can save our lives. For this reason, I believe that the book might also be interesting for decision makers. Just in case, and to take better decisions based on PC and AHP.

1.2   AHP basics

The beginning of the *Analytic Hierarchy Process,* abbreviated to AHP, is assumed to be the work *"A scaling method for priorities in hierarchical structures"* [155] published by *Thomas Saaty* in 1977. Interestingly, the name AHP does not appear in the work itself. However, in the next article [57], submitted to The European Journal of Operational Research in the fall of 1977, *Saaty* used the name of the new method explicitly in the title. From this point of time, the name of the new method entered into the scene of MCDM. Although in his seminal work [155] *Saaty* does not use the name AHP, he defines almost all the important principles of the method. In particular, he addresses such tough issues as a priority-deriving method, the inconsistency of pairwise comparisons, inconsistency measurements, measurement scale or hierarchy. As a result, people all over the world received a new, relatively simple, well defined and complete decision-making method helping them to deal with many alternatives and criteria. Due to its simplicity and intuitive mathematical foundations, and as some argue excellent marketing and business intuition of *Saaty*, for years AHP gained crowds of supporters and loyal users. Over time, it turned out that some solutions proposed in [155] can be questioned due to their sub-optimality or doubtful mathematical basis. However, despite the fact that the building of AHP erected by *Saaty* has been crushed many times under criticism (very often the right criticism), AHP survived and is still probably one of the most popular MCDM methods in the world.

AHP is not only a mathematical recipe allowing the user to insert some data to the input and get the ready-to-use solution on the output. In [155] we can see *"A fundamental problem of decision theory is how to derive weights for a set of activities according to importance. (...) This is a process of multiple criterion decision making (...)".* It is clear that *Saaty* perceived the method he created as the process of transforming the experts' knowledge into the best possible recommendation and/or decision. The elements of this process are humans, mathematical methods, and recently more and more appropriate software and dedicated devices. Cooperation of all these elements over time results in the rendering of the final recommendation. Thus, AHP can be analyzed from many different angles, depending on the component of the method that is important to the researcher. The heart of the method, however, is the mathematical theory that we intend to unveil in this book.

1.2.1   Actors

AHP is an expert method. This means that the source of the decision data and an important element of the method itself are the *experts*. These are the people whose individual judgments are used for the final recommendation. Although experts are very important to the AHP procedure, they are not the only actors involved in decision-making processes (Fig. 1.1). Another

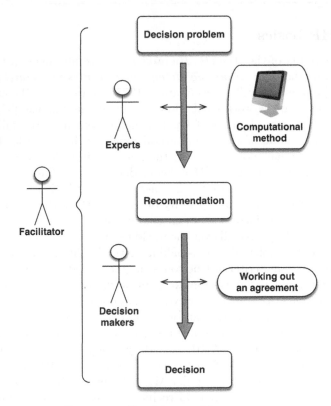

Figure 1.1: Decision-making process

important player involved in the process is a *facilitator*[1] or a *supervisor*. The facilitator is a person responsible for the smooth conduct of the process from its very beginning identifying the decision problem – to the very end, including

---

[1]For example, we may find a facilitator in the PROMETHEE method. See [61, p. 183].

preparing the recommendation and rendering the final decision. This is a person who knows the procedure. All the other parties involved trust that the facilitator will not unfairly interfere in the process and the final outcome will actually reflect the actual experts' opinions. The last group of actors involved in the process are the decision-makers. They have the right and sufficient power to make the final resolution. For this purpose, they use the recommendation prepared by the experts and the facilitator. Based on their professional knowledge, intuition, deep understanding of the problem and, first of all, common sense, they turn the recommendation into a decision. Hence, the recommendation is not the same as the outcome of the decision process. It is possible that the final decision differs from the recommendation, but if this happens the decision-makers must be able to give a good and convincing explanation as to why it is so.

The experts, facilitator(s) and decision-makers are generally distinct people. However, these roles can sometimes be combined together for various reasons. For example, experts may also play the role of decision-makers. Probably for this reason, decision-makers are quite often identified with experts in the literature. Of course, one of the experts can also act as a facilitator. From our perspective, the most important role will be the latter. He or she will be the one who knows all the details of the computational method – their strengths and weaknesses.

1.2.2   Alternatives and criteria

AHP is sometimes called a ranking method. That is because AHP is good at prioritizing different things, objects or concepts. We will call all these things *alternatives* and denote by a small letter a. The overall universe of all the possible alternatives is denoted by the capital letter A. Thus every single alternative belongs to A, which might be written in a more formal way as $a \in A$. Essentially, we do not want to deal with all the possible alternatives that exist or which we can come up with. Thus, in every particular decision problem there will be a finite set of alternatives:

$$A = \{a_1, \ldots, a_n\}$$

to which we will limit our attention. The purpose of AHP is to help introduce order into A, and thus provide information to the decision-makers about the alternative that is most beneficial. What can be an alternative? Basically anything, from political decisions through cars to a set of candidates for managerial positions. A good selection of examples can be found in the *Saaty* and *Vargas* book *"Models, Methods and Concepts & Applications of the Analytic Hierarchy Process"* [169].

Even such a well-defined and relatively simple to describe item as a chair might be considered in terms of the many different features it has, like color, height, stability (safety), shape, usability, durability and so on. Thus, being in the middle of a furniture store, we very often can not determine which chair a_1 or a_2 is better (whatever it means to us), but when it comes to deciding which color we prefer we usually do not have problems with that. This simple example shows that it is much easier to assess an alternative with respect to the given feature than the alternative in general. To capture the variety of different attributes that alternatives may possess, AHP, as well as many other decision-making methods, use the notion of a criterion. Hence, the methods that allow the user to rank alternatives with respect to many different attributes are called MCDM methods i.e. multiple-criteria decision-making methods.

1.2.3   Comparing alternatives in pairs

What can be a criterion? Like before, a criterion can be almost anything, however, this time it will be convenient to split criteria into tangible and intangible ones. Tangible criteria are those for which we have a yardstick. A kind of measure or a tool which allows us to accurately determine appropriate magnitude, like the size, length, height, temperature, price and so on. Intangible criteria are usually abstract concepts like beauty, usefulness or potential. For the latter group of criteria, there are neither methods nor units that allow us to determine their values. It does not mean, however, that we cannot express an opinion that the alternative a_1 (let's say that a_1 means an oil painting of one famous painter) is more beautiful than the painting a_2. In other words, although we can not measure some features of alternatives, we are still able to compare them. This observation underlies different decision methods based on comparing alternatives in pairs [48, 41, 43, 42]. One of them is AHP. The role of an expert in all those methods is to compare all the alternatives in pairs with respect to the given criterion or criteria. Thus, the input to the method is a set (or sets) of pairwise comparisons, while the output is the ranking vector consistent with the revealed preferences of experts. The result of a single comparison of the two alternatives is the number defining how much the first alternative is better (more preferred, more important) than the second one with respect to the given criterion. In the literature, it is very often assumed that this is a real positive number. For this reason, in the book we also accept that the result of a comparison of two alternatives a_i and a_j is a $c_{ij} \in \mathbb{R}_+$. However, as we explain in Section 1.2.4, in the "traditional" AHP it is a positive, rational number.

For convenience, the set of pairwise comparisons for the given set of alternatives $A = \{a_1, \ldots, a_n\}$ is written in the form of *the pairwise comparisons (PC) matrix*:

$$C = \begin{pmatrix} 1 & c_{12} & \cdots & c_{1n} \\ c_{21} & 1 & \cdots & c_{2n} \\ \vdots & \vdots & \ddots & \vdots \\ c_{n1} & c_{n2} & \cdots & 1 \end{pmatrix}. \tag{1.1}$$

As we can see, the matrix C on its diagonal has only ones. This stems from the fact that the entries in the form of c_{ii} correspond to the comparisons of alternatives with themselves. In the pair (a_i, a_i) both alternatives are equally preferred, thus the value of $c_{ii} = 1$ for $i = 1, \ldots, n$. In AHP, the PC matrix C has one more important property. This is reciprocity. It is said that the matrix is reciprocal if for its entries $c_{ij} = 1/c_{ji}$. The reciprocity models the principle that it does not matter whether we compare a_i to a_j or reversely a_j to a_i. In other words, if a_i turns out to be two times more preferred than a_j, then of course a_j must be two times less preferred than a_i. Due to reciprocity, we may rewrite the matrix C as follows:

$$
C = \begin{pmatrix}
1 & c_{12} & \cdots & c_{1n} \\
\frac{1}{c_{12}} & 1 & \cdots & c_{2n} \\
\vdots & \vdots & \ddots & \vdots \\
\frac{1}{c_{1n}} & \frac{1}{c_{2n}} & \cdots & 1
\end{pmatrix}.
$$

For the same reason, it is very often assumed that to specify the PC matrix it is enough to provide entries above its main diagonal.

Let us go back for a moment to the tangible criteria. One may ask whether AHP is suitable for them? The answer is not immediate. Of course, if we compare two alternatives with respect to a tangible feature, and we know its exact values, it is enough to make a simple division to get a result of comparison. In such a case, we do not need an expert, but a good calculator. However, even when the criterion is tangible, we are sometimes not able to measure their actual values. For example, when walking along a mountain trail, if we do not have a map but we want to know how far point A is from point B, we have to ask a more experienced tourist for the distance estimation. So, even though that distance can be perfectly measured, we have to rely on the expert's opinion. On the other hand, we may sometimes be interested in the relative perception of some magnitude. In this case, the actual value is not relevant to us, but the respondent's subjective perception of that magnitude is. For this reason, if we ask our neighbors how they rate the price increase in the last year, we will get information about how the price hike affected them rather than the actual value of the inflation rate. The decision how to handle tangible criteria usually depends on a facilitator. It is worth noting that *Saaty* tends to recommend using pairwise comparisons always for both tangible and intangible criteria [169, p. 64].

1.2.4 Scale

In the book *"Measurement Theory with Applications to Decisionmaking, Utility, and the Social Sciences"* *Fred S. Roberts* defines a measurement scale as a mapping[2] whose values have some specific properties [152]. One of the

---

[2]To be strict, the measurement scale is a triple $(\mathfrak{A}, \mathfrak{B}, f)$ composed of two relational systems $\mathfrak{A} = (A, H, \circ)$, $\mathfrak{B} = (\mathbb{R}_+, >, +)$ and a homomorphism f [152, pp. 51–53, p. 64]. However, for the purpose of this somewhat informal introduction to AHP, we do not need to formally define this notion.

required properties might be preserving the order (relation) between values. Hence, if the scale $f_o : X \to \mathbb{R}$ takes two terms $x_1, x_2 \in X$ as its arguments and $f_o(x_1) < f_o(x_2)$ then we may also expect that for every monotonic function $g : \mathbb{R} \to \mathbb{R}$ it holds that $g \circ f_o(x_1) < g \circ f_o(x_2)$. In other words, this scale, called the ordinal scale [152, p. 64], allows us to perform any transformation of its values as long as the transformation preserves the order of elements. An example is the scale of air pollution used in a number of cities. It assigns the number 1 to clear air, 2 to low pollution, 3 to moderate pollution, 4 to high pollution, and 5 to extreme pollution. However, we may adopt any numbers that preserve the order as the levels of pollution. For instance, $1, 3, 5, 7, 9$ are equally good.

The other important mapping is the ratio scale $f_r : X \to \mathbb{R}$. The ratio scale allow us to perform similarity transformations, i.e., any $g : \mathbb{R} \to \mathbb{R}$ in the form $g(x) = \alpha x$ where $\alpha \in \mathbb{R}_+$ used as a transformation of f_r makes sense. The example of the ratio scales are the length with its unit conversions between centimeters and inches, weight, pressure and so on.

The set X can take different forms. In particular, it can be a Cartesian product of some other relational sets. For the purpose of comparing alternatives in AHP, *Saaty* defined a scale that maps the Cartesian product of two sets of *verbal expressions* to the subset of rational numbers $\{1/9, 1/8, \ldots, 1/2, 1, 2, \ldots, 9\}$. This scale was named by *Saaty* as a fundamental scale [156]. It is very often presented in the form of a table (Table 1.1).

The fundamental scale was introduced as a method to help experts compare alternatives. An expert using the scale needs to decide whether a_1 is equal, weak, essential, demonstrated or absolute more preferred than a_2, and if for example he/she decides that a_1 is essentially more important than a_2 then, as a result of the comparisons, he/she adopts $c_{12} = 5$. When, reversely, he/she decides that a_2 is essentially more important than a_1 the comparison c_{12} takes the value $1/5$. It is worth noting that as every PC matrix entry comes from the set $\{1/9, 1/8, \ldots, 1/2, 1, 2, \ldots, 9\}$ then the PC matrices contain only rational entries.

In fact, in his theoretical works on AHP, *Saaty* extends the notion of the fundamental scale to the mapping[3] $f : T \times T \to \mathbb{R}_+$ where T is any ordered set of verbal expressions [159, 17, 167]. Despite this extension, the fundamental scale as presented in Table 1.1 has become widespread and very popular. In fact, its popularity in this shape was influenced by *Saaty* himself, who on many occasions presented it in this simplified form [196, 162].

The fundamental scale still raises a lot of controversy. One of them is connected with the fact that people may have various numerical interpretations of the same verbal expressions. This may result in the lower accuracy of the

---

[3] *Saaty* instead of f uses the symbol P_C as every pair $(a_i, a_j) \in T \times T$ corresponds to a pairwise comparison of the i-th and j-th alternative [17, 167].

Verbal expression	Value of intensity	Explanation
equal importance	1	two alternatives are equally important
moderate importance of one over another	3	experience and judgment slightly favor one alternative over another
essential or *strong* importance	5	experience and judgment strongly favor one alternative over another
demonstrated importance	7	an alternative is strongly favored and its dominance demonstrated in practice
absolute importance	9	the evidence favoring one alternative over another is the highest possible
intermediate values	2,4,6,8	if predominance of one alternative over another fits in between the scale degrees defined above

Table 1.1: Fundamental scale

results [84]. Another, probably more serious, one concerns the nature of this scale. Obviously, it does not fit into the classification given by *Roberts*[4] [15, p. 80]. This raises the question of the type of operations that can be performed on this scale, especially in the light of the fact that, on the one hand, we have a set of verbal expressions with an ordinal nature, while on the other hand there is $\mathbb{R}_+$ whose elements have a cardinal meaning. Some authors argue that the rankings obtained by using AHP may not be meaningful[5] [55, 15]. These theoretical objections, however, do not interfere with practice. The "nine point" fundamental scale proposed by *Saaty,* as well as the other scales, are very popular among practitioners. Some of them will be discussed later on in Chapter 5.

In order to not deal with the correctness of the fundamental scale for most of the book, we will assume that the expert judgments have a cardinal meaning. Thus, if for example an expert decides that c_{12} should be 5, this will mean that the alternative a_1 is exactly five times more preferred than a_2 (not "essentially more preferred," whatever in fact it means). All exceptions to this rule will be clearly indicated in the text.

---

[4]The problem with the description of AHP is not the only shortcoming of the Roberts theory. *Barzilai* also indicates its other limitations [15, p. 80].

[5]*Dyer's* critical comments [55, 54] met with *Saaty's* prompt reply [161].

1.2.5 Prioritization

Prioritization is a computational procedure which takes all the pairwise comparisons as the input and produces the ranking vector as the output (Fig. 1.2). The set of pairwise comparisons, usually in the form of a PC matrix, is prepared by experts.

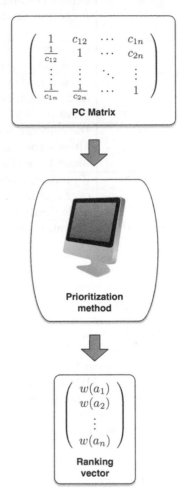

Figure 1.2: Prioritization procedure

Then the PC matrix is taken by a facilitator who checks the data quality and runs the ranking procedure. As a result, a ranking vector is created. This is a vector of weights. Let us define the weight function as $w : A \to \mathbb{R}_+$ assigning to every considered alternative $a_i \in A$ a real positive value $w(a_i) \in \mathbb{R}_+$. As $A = \{a_1, \ldots, a_n\}$ is discrete and finite, it is convenient to represent w in the form of a vector:

$$
w = \begin{pmatrix} w(a_1) \\ w(a_2) \\ \vdots \\ w(a_n) \end{pmatrix}.
$$

The meaning of w is twofold: qualitative and quantitative. On the one hand, w introduces the absolute order into A. Thus, whenever $w(a_i) > w(a_j)$ we accept that a_i is more preferred (or more important) than a_j. Of course, if $w(a_i) = w(a_j)$ then a_i is considered as equally preferred (or equally important) to a_j. In practice, wherever decision makers get a recommendation in the form of w, they look for its maximum and if a_i maximizes w, then they adopt a_i as their decision. Sometimes it is possible that more than one alternative is maximal. In such a case, the decision makers have to assume some tie breaking rule. Of course, decision makers may also adopt as their choice the alternative corresponding to the second value, the third maximal and so on, in w as these alternatives represent the second, third and next best decisions they can take.

On the other hand, w also carries cardinal information as to the importance of alternatives. Therefore, if for instance $w(a_i)/w(a_j) = 2$ then the alternative a_i is considered to be two times more worthy of attention than the alternative a_j. It may also have a practical dimension. Let us consider an example, somewhat theoretical but not difficult to imagine in practice, of n projects $a_1, \ldots, a_n$ competing for a share in sponsorship funds. The amount of money to be divided is known in advance and equals N. As a result of the pairwise comparisons and the prioritization procedure, the ranking vector $w = [w(a_1), \ldots, w(a_n)]^T$ was prepared. Since it is reasonable to expect that the share of a_i in funds should be proportional to the score it got in the ranking thus, providing that $\sum_{i=1}^n w(a_i) = 1$, the amount of money a_i will receive is $w(a_i) \cdot N$. In such a case, the most important information for teams is not who the winner is but what their ranking score is.

Despite the quantitative information contained in the ranking vector in the prevailing number of cases, AHP's output is interpreted qualitatively. More important is the order determined by the weights than the weights themselves.

Different prioritization methods used in the context of AHP will be throughly discussed in Chapter 3.

1.2.6   Inconsistency

The assessments of experts are the input data for the ranking procedure (Fig. 1.1). As ordinary people, they can be tired or sleepy, sad or upset. They can make mistakes or be imprecise in their judgments. As a result, it is possible that in a PC matrix C there are entries c_{ij}, c_{jk} and c_{ik} such that $c_{ij}, c_{jk} > 1$ and $c_{ik} < 1$. This means that the alternative a_i is more preferred than a_j, a_j is more preferred than a_k but a_i is less preferred than a_k. Because it is

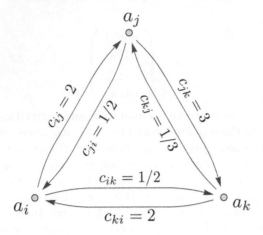

Figure 1.3: Inconsistent triad a_i, a_j, and a_k as a graph

widely accepted that the relation of being "more preferred" is transitive[6], such a situation indicates ordinal inconsistency in the expert's preferences. In fact, we may expect even more from the matrix C. Since in AHP the pairwise comparisons have a quantitative meaning, we can demand that $c_{ij} \cdot c_{jk} = c_{ik}$. For example, if according to the experts a_i is two times more preferred than a_j, which means that $c_{ij} = 2$, and a_j is three times more preferred than a_k ($c_{jk} = 3$), then it seems natural to expect that $c_{ik} = 6$. As before, this condition may not be met in practice. The situation in which there are at least three entries in C such that $c_{ij} \cdot c_{jk} \neq c_{ik}$ indicates the occurrence of the *cardinal inconsistency*. An example of an ordinal and cardinal inconsistent triad is shown in Figure 1.3.

The preferential inconsistency is an inherent phenomenon accompanying the preferential relation. The inevitability of this phenomenon is highlighted by von *Neumann* and *Morgenstern* in their seminal book *"Theory of Games and Economic Behavior"* [135, pp. 37–39]. Inconsistency may indicate that the expert was either unreliable or unsure, hence, following the popular saying, garbage in – garbage out, the outcome of the ranking procedure may also be questionable. For these reasons, AHP also has to deal with the inconsistency. It addresses the problem by providing the robust prioritization method which returns the correct and admissible ranking vector, even when the input data are inconsistent. On the other hand, *Saaty* proposed a consistency index – the mathematical function which allows the users to determine the extent to which the PC matrix is inconsistent. Thus, knowing the consistency level for

---

[6]In preference theory, this property is called the axiom of transitivity [60, p. 12]. It also covers the situation when experts are indifferent between two alternatives. The other important assumption is called the axiom of completeness that for every two alternatives a_i and a_j either a_i is more preferred than a_j or a_j is more preferred than a_i or both are equally preferred.

the given PC matrix, the decision makers may decide whether to accept the ranking or reject it.

There is an ongoing debate among the people involved in the PC method about the concept of inconsistency and ways of measuring it. Thus, there are many inconsistency indices [33, 104]. The role of inconsistency in the AHP method [64], as well as issues such as whether inconsistency is desirable or not, can also be debated. Saaty, indicating its significance, wrote *"(...)* *inconsistency itself is important because without it new knowledge that changes* *preference cannot be admitted. Assuming that all knowledge should be consis-* *tent contradicts experience, which requires continued revision of understand-* *ing"* [165, p. 201]. Due to its importance, this book also deals with the preferential inconsistency. Thus, a more detailed analysis of this notion as well as its properties is provided in Chapter 6.

1.2.7 Hierarchy

AHP, as was mentioned at the beginning of the chapter, is a MCDM method. Thus, wherever we compare one alternative against the other, we do that in the context of some criterion or feature. Of course, sometimes alternatives are such simple objects that there is no sense considering them in the context of many criteria. Another time, we may only care about one particular characteristic, so all the other aspects of the considered objects are, in fact, irrelevant. In all these cases, it is enough to construct one PC matrix containing all indispensable comparisons, then use a prioritization method and obtain the ranking. However, if the alternatives we are interested in have more than one important characteristic, we need to take into account all of them when creating the ranking. Thus, in such a case we have to create rankings of all the alternatives for every single criterion, then combine all these partial results together, then form one global ranking vector. Of course, different criteria contribute to the final ranking differently. Some of the characteristics we consider are more important than others. For this reason, we may expect that if the alternative a_i wins against a_j with respect to some important criterion, then it matters more than the fact that a_i loses to a_j with respect to some minor criterion. To decide the importance of different criteria, we need to rank them. Of course, the recommended way to do that is comparing alternatives in pairs and deriving the weight vector by using any prioritization method. This is how the hierarchy of rankings begins. At the lowest level of the hierarchy, we rank alternatives with respect to criteria. Higher up we rank criteria with respect to their importance. Even higher we may introduce the higher level criteria (groups of criteria) and so on. In order to imagine how this works, let us consider n alternatives in the form $a_{i,l}$ where i enumerates the subsequent alternatives and l refers to the level of a hierarchy[7]:

$$A_1 = \{a_{1,1}, \ldots, a_{n,1}\}.$$

They have m important characteristics, thus we have to consider them with respect to m criteria. Let us denote these criteria as:

$$A_2 = \{a_{1,2}, \ldots, a_{m,2}\}.$$

Thus, in the first step of our procedure, we ask experts to prepare m PC matrices $C_{q,r} = [c_{i,j,nc,l}]$ where i,j refers to the numbers of compared alternatives and nc, l denotes the number of criterion and the level in the hierarchy, respectively. In practice, wherever it is possible, the pair of additional indices nc, l will be omitted.

$$C_{1,1} = \begin{pmatrix} 1 & c_{1,2,1,1} & \cdots & c_{1,n,1,1} \\ \frac{1}{c_{1,2,1,1}} & 1 & \cdots & c_{2,n,1,1} \\ \vdots & \vdots & \ddots & \vdots \\ \frac{1}{c_{1,n,1,1}} & \frac{1}{c_{2,n,1,1}} & \cdots & 1 \end{pmatrix}, \ldots,$$

$$C_{m,1} = \begin{pmatrix} 1 & c_{1,2,m,1} & \cdots & c_{1,n,m,1} \\ \frac{1}{c_{1,2,,m,1}} & 1 & \cdots & c_{2,n,,m,1} \\ \vdots & \vdots & \ddots & \vdots \\ \frac{1}{c_{1,n,,m,1}} & \frac{1}{c_{2,n,,m,1}} & \cdots & 1 \end{pmatrix}$$

Then providing that all the matrices are consistent enough, we calculate m ranking vectors $w_{nc,l}$ on the first level of a hierarchy:

$$w_{1,1} = \begin{pmatrix} w_{1,1}(a_{1,1}) \\ w_{1,1}(a_{2,1}) \\ \vdots \\ w_{1,1}(a_{n,1}) \end{pmatrix}, \ldots, w_{m,1} = \begin{pmatrix} w_{m,1}(a_{1,1}) \\ w_{m,1}(a_{2,1}) \\ \vdots \\ w_{m,1}(a_{n,1}) \end{pmatrix}$$

Next, to determine the importance of each criterion, experts are asked to prepare the PC matrix of criteria.

$$C_{1,2} = \begin{pmatrix} 1 & c_{1,2,1,2} & \cdots & c_{1,m,1,2} \\ \frac{1}{c_{1,2,1,2}} & 1 & \cdots & c_{2,m,1,2} \\ \vdots & \vdots & \ddots & \vdots \\ \frac{1}{c_{1,m,1,2}} & \frac{1}{c_{2,m,1,2}} & \cdots & 1 \end{pmatrix}$$

This leads to the criteria priority vector on the second level of a hierarchy:

---

[7]Considering the following scheme we slightly overuse the number of indices. This abuse, however, allows us to express a general formula for calculating global ranking in AHP.

$$w_{1,2} = \begin{pmatrix} w_{1,2}(a_{1,2}) \\ w_{1,2}(a_{2,2}) \\ \vdots \\ w_{1,2}(a_{m,2}) \end{pmatrix}$$

In the last step, we have all these results combined together into one overall ranking (Fig. 1.4). We may observe that the first alternative a_1 received the rank score $w_{1,1}(a_{1,1})$ according to the first criterion, while the first criterion $a_{1,2}$ itself has the weight $w_{1,2}(a_{1,2})$. In general, the weight of a_i with respect to the j-th criterion is $w_{j,1}(a_{i,1})$ and the weight of the j-th criterion is $w_{1,2}(a_{j,2})$.

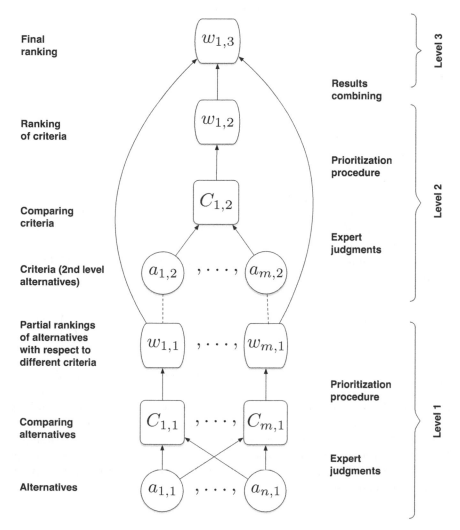

Figure 1.4: Three-level AHP computation model

Thus, the importance of the j-th weight of the i-th alternative, i.e., $w_{j,1}(a_{i,1})$ is determined by $w_{1,2}(a_{j,2})$. For this reason, the final weight for a_i is given as the sum of m components:

$$w_{1,2}(a_{1,2}) \cdot w_{1,1}(a_{i,1}) + \ldots + w_{1,2}(a_{m,2}) \cdot w_{m,1}(a_{i,1}).$$

We may denote them as the ranking vector on the third level, i.e.,

$$w_{1,3} = \begin{pmatrix} w_{1,3}(a_{1,1}) \\ w_{1,3}(a_{2,1}) \\ \vdots \\ w_{1,3}(a_{m,1}) \end{pmatrix},$$

where

$$w_{1,3}(a_{i,1}) = \sum_{j=1}^{m} w_{1,2}(a_{j,2}) \cdot w_{j,1}(a_{i,1}). \tag{1.2}$$

To compute $w_{1,3}$ we need m vectors $w_{1,1}, \ldots w_{m,1}$ from the first level and one second level vector $w_{1,2}$ (Fig. 1.4).

It is possible, however, that for various reasons we want to group the criteria. In such a case, on the second level of a hierarchy there are m_3 groups of criteria. Criteria within the same group are compared against each other and the group ranking of criteria is created. At the next level of the hierarchy, here the third level, groups of the criteria are compared against each other. Thus, the weight of an alternative with respect to the given criterion is scaled by the weight of the criterion and the weight of the group of criteria to which the criterion belongs. Therefore, for a 4-level hierarchy, the final weight of a_i is given by the expression

$$w_{1,4}(a_{i,1}) = \sum_{j_3=1}^{m_3} w_{1,3}(a_{j_3,3}) \left(\sum_{j_2=(m_{j_3-1})+1}^{m_{j_3}} w_{j_3,2}(a_{j_2,2}) \cdot w_{j_2,1}(a_{i,1}) \right), \tag{1.3}$$

where there are m_3 groups of the criteria on the third level of the hierarchy, every j_3-th group of the criteria $a_{j_3,3}$ contains exactly m_{j_3} criteria and the ranking of the groups of the criteria is represented by $w_{1,3}$. The final ranking for the alternatives $a_{1,1}, \ldots, a_{n,1}$ is given as the vector $w_{1,4}$ (Fig. 1.5).

Of course, the hierarchy can be arbitrarily expanded. One can imagine that the criteria are grouped, those groups form larger groups and so on. For example, for r levels of the hierarchy, the priority of the i-th alternative is given by the expression:

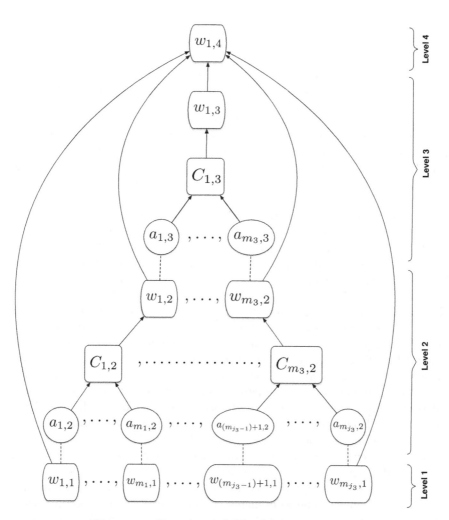

Figure 1.5: Four-level AHP computation model

$$w_{1,r}(a_{i,1}) = \sum_{j_{r-1}=1}^{m_{r-1}} w_{1,r-1}(a_{j_{r-1},r-1})$$

$$\left(\sum_{j_{r-2}=(m_{j_{r-1}-1})+1}^{m_{j_{r-1}}} w_{j_{r-1},r-2}(a_{j_{r-2},r-2}) \right.$$

$$\left. \left(\cdots \left(\sum_{j_2=(m_{j_3}-1)+1}^{m_{j_3}} w_{j_3,2}(a_{j_2,2}) \cdot w_{j_2,1}(a_{i,1}) \right) \right) \cdots \right)$$

One may ask about the purpose of creating the groups of criteria and hence the extension of the hierarchy. Could it not be confined to only three levels: the bottom level for alternatives, middle one for criteria and the upper one for assembling them together? From a technical point of view, this could be done, however, there are two arguments for not doing so. One of the reasons is that it is easier to compare things and concepts when they are similar than when they are completely different. Hence, if among the criteria there is a group of similar concepts, it makes sense to limit the comparisons to this group and not to compare them to others which are not similar to them. This observation underlies the rule adopted by *Saaty* as the axiom of homogeneity [159, 167]. According to the principle *"as the mind tends to make large errors in comparing widely disparate elements,"* experts should compare only those "elements" which are to some extent similar. In other words, comparisons should be performed only within homogeneous groups. The second argument for a hierarchy larger than three has a practical nature. When the number of criteria is large, it might be difficult to compare all of them in pairs. For the purpose of AHP, *Saaty* adopted *"the magical number seven plus or minus two"* as the upper limit of the number of items compared against each other: *"each cluster must not contain more than a few elements: about seven"* [165, 168]. The idea of seven he gets from *Miller* [132], who argued that this is the human capacity for processing information. Interestingly, *Saaty* gives the *magic number seven*, specifically *seven plus* two as the reason why the fundamental scale (Sec. 1.2.4) is composed of nine successive values (seven plus two make nine) [156, p. 151]. Whether or not we accept this argument, it is obvious that the number of comparisons we have to perform increases in a geometric progression with respect to the number of compared items. Hence, it is reasonable to keep the number of directly compared alternatives or criteria low. From this point of view, it is clear that the concept of a hierarchy is crucial to handle decision problems where the number of criteria is large – larger than the magic number *"seven plus two"* or any other reasonable number we accept as the limit for the size of the set of mutually compared alternatives.

1.3   Neat examples

A careful reader might have noticed that the previous considerations (Section 1.2) are missing two important details. The first one is the lack of a prioritization method, the second one is the lack of an inconsistency index. This lack is intentional. Over the years, apart from the methods originally proposed by *Saaty,* a number of other solutions have emerged. They concern both: the calculation of priorities and the measurement of inconsistency. Their authors argue that they are at least as good as the *Saaty* methods and very often even better than the original solutions. These claims are frequently supported by strong evidence. Thus, writing this chapter, we did not want to suggest to the reader in advance any particular way or a method. Some prioritization methods, including the one proposed by *Saaty,* will be discussed in Chapter 3. Inconsistency indices are considered in Chapter 6. However, it is difficult to discuss numerical examples without using numerical methods. Thus, for the purpose of the following two examples, a prioritization procedure as well as inconsistency index will be treated like a black box. In other words, we will use some PC matrices as the input and get the ranking vector and the inconsistency score as the output. In fact, we will use *Saaty's* methods for the calculations, thus everybody who will read Chapters 3 and 6 may go back and recalculate both examples by themselves.

1.3.1   Satisfaction with house

The problem discussed here concerns purchasing a house for a family [165]. After browsing the offers, three homes have been selected for further consideration. During a meeting, the members of the family identified the eight criteria that are important to them (Fig. 1.6). These are:

1. price – affordability in terms of the level of the monthly loan installment, creditworthiness of the family,

2. size of house – number of bedrooms, number of bathrooms, overall area of the house, size of rooms,

3. transportation – convenience and proximity of means of public transportation, like bus or subway,

4. neighborhood – ambient noise, physical condition of surrounding buildings, insurance conditions, taxes, security,

5. age of house – number of years from completion of construction to now,

6. yard space – how much space is available around the house, including parking space,

7. facilities – visual monitoring, alarm system, intercom, air conditioning, refrigerator, dishwasher, built-in furniture,

8. general condition – condition of walls, floors, ceilings, wiring, electricity network, overall cleanliness.

Then, they compared all three houses with respect to each criterion. As a result, the following eight matrices were created:

$$C_1 = \begin{pmatrix} 1 & \frac{1}{7} & \frac{1}{5} \\ 7 & 1 & 3 \\ 5 & \frac{1}{3} & 1 \end{pmatrix}, \quad C_2 = \begin{pmatrix} 1 & 5 & 9 \\ \frac{1}{5} & 1 & 4 \\ \frac{1}{9} & \frac{1}{4} & 1 \end{pmatrix},$$

$$C_3 = \begin{pmatrix} 1 & 4 & \frac{1}{5} \\ \frac{1}{4} & 1 & \frac{1}{9} \\ 5 & 9 & 1 \end{pmatrix}, \quad C_4 = \begin{pmatrix} 1 & 9 & 4 \\ \frac{1}{9} & 1 & \frac{1}{4} \\ \frac{1}{4} & 4 & 1 \end{pmatrix},$$

$$C_5 = \begin{pmatrix} 1 & 1 & 1 \\ 1 & 1 & 1 \\ 1 & 1 & 1 \end{pmatrix}, \quad C_6 = \begin{pmatrix} 1 & 6 & 4 \\ \frac{1}{6} & 1 & \frac{1}{3} \\ \frac{1}{4} & 3 & 1 \end{pmatrix},$$

$$C_7 = \begin{pmatrix} 1 & 9 & 6 \\ \frac{1}{9} & 1 & \frac{1}{3} \\ \frac{1}{6} & 3 & 1 \end{pmatrix}, \quad C_8 = \begin{pmatrix} 1 & \frac{1}{2} & \frac{1}{2} \\ 2 & 1 & 1 \\ 2 & 1 & 1 \end{pmatrix},$$

where C_1 corresponds to the first criterion, i.e., price, C_2 represents size of house, and so on.

Figure 1.6: Schema of the hierarchy

As a result of prioritization, the following priority vectors have been calculated[8]:

$$w_1 = \begin{pmatrix} 0.072 \\ 0.649 \\ 0.279 \end{pmatrix}, \quad w_2 = \begin{pmatrix} 0.743 \\ 0.194 \\ 0.063 \end{pmatrix}, \quad w_3 = \begin{pmatrix} 0.194 \\ 0.063 \\ 0.743 \end{pmatrix},$$

$$w_4 = \begin{pmatrix} 0.717 \\ 0.0658 \\ 0.217 \end{pmatrix}, \quad w_5 = \begin{pmatrix} 0.333 \\ 0.333 \\ 0.333 \end{pmatrix}, \quad w_6 = \begin{pmatrix} 0.691 \\ 0.0914 \\ 0.217 \end{pmatrix},$$

---

[8]In this and the following example, we use the so-called eigenvalue priority-deriving method. It will be defined and explained in one of the following chapters.

$$w_7 = \begin{pmatrix} 0.77 \\ 0.068 \\ 0.162 \end{pmatrix}, \quad w_8 = \begin{pmatrix} 0.2 \\ 0.4 \\ 0.4 \end{pmatrix}.$$

The vector of weights w_1 is derived from C_1, w_2 comes from C_2 and so on. For example, according to price (first criterion), the most attractive seems to be the second house (weight 0.649), then house number three with the weight 0.279, and at the end of the list is house no. 1 with the weight 0.072.

In the next step, the experts, in this case family members, have to assess the importance of the eight criteria. For this purpose, they create the second level PC matrix $C_{1,2}$ containing comparisons of all the criteria $a_{1,2}, \ldots, a_{8,2}$:

$$C_{1,2} = \begin{pmatrix}
1 & 4 & 7 & 5 & 8 & 6 & 6 & 2 \\
\frac{1}{4} & 1 & 5 & 3 & 7 & 6 & 6 & \frac{1}{3} \\
\frac{1}{7} & \frac{1}{5} & 1 & \frac{1}{3} & 5 & 3 & 3 & \frac{1}{5} \\
\frac{1}{5} & \frac{1}{3} & 3 & 1 & 6 & 3 & 4 & \frac{1}{2} \\
\frac{1}{8} & \frac{1}{7} & \frac{1}{5} & \frac{1}{6} & 1 & \frac{1}{3} & \frac{1}{4} & \frac{1}{7} \\
\frac{1}{6} & \frac{1}{6} & \frac{1}{3} & \frac{1}{3} & 3 & 1 & \frac{1}{2} & \frac{1}{5} \\
\frac{1}{6} & \frac{1}{6} & \frac{1}{3} & \frac{1}{4} & 4 & 2 & 1 & \frac{1}{5} \\
\frac{1}{2} & 3 & 5 & 2 & 7 & 5 & 5 & 1
\end{pmatrix}. \tag{1.4}$$

The priority vector derived from $C_{1,2}$ is

$$w_{1,2} = \begin{pmatrix} 0.345 \\ 0.175 \\ 0.062 \\ 0.103 \\ 0.019 \\ 0.034 \\ 0.041 \\ 0.22 \end{pmatrix}.$$

The last step is to merge all the eight priority vectors $w_1, \ldots, w_8$ into one overall ranking using the ranking of the criteria $w_{1,2}$. Following (1.2) the overall weight of the i-th house is

$$w(a_i) = \sum_{j=1}^{8} w_{1,2}(a_{j,2}) \cdot w_j(a_i) \quad \text{for } i = 1, 2, 3. \tag{1.5}$$

Thus, for example, for $i = 1$ we get

$$w(a_1) = 0.345 \cdot 0.072 + 0.175 \cdot 0.743 + \ldots + 0.22 \cdot 0.2 = 0.346347.$$

In a similar way, we can compute the weights for a_2 and a_3 which are $w(a_2) = 0.36914$ and $w(a_3) = 0.284513$ respectively. As the second house (alternative) has the highest weight, the family decided to choose house number two.

The equation (1.5) can also be written in the matrix form:

$$w = [w_1, \ldots, w_8] \cdot w_{1,2}$$

where $[w_1, \ldots, w_8]$ means the matrix 8×3 in which every i-th column corresponds to the vector w_i. In this case, the equation takes the form:

$$\begin{pmatrix} 0.072 & 0.743 & 0.194 & 0.717 & 0.333 & 0.691 & 0.77 & 0.2 \\ 0.649 & 0.194 & 0.063 & 0.066 & 0.333 & 0.091 & 0.068 & 0.4 \\ 0.279 & 0.063 & 0.743 & 0.217 & 0.333 & 0.218 & 0.162 & 0.4 \end{pmatrix} \cdot \begin{pmatrix} 0.345 \\ 0.175 \\ 0.062 \\ 0.103 \\ 0.019 \\ 0.034 \\ 0.041 \\ 0.22 \end{pmatrix} =$$

$$= \begin{pmatrix} 0.346 \\ 0.369 \\ 0.284 \end{pmatrix}$$

1.3.2 Car selection

Another very popular application of AHP is the choice of a car. Various examples of car selection problems can be found in the literature [36, 6, 139] and the Internet[9]. Because this problem is somewhat natural for AHP (as it combines tangible and intangible criteria), we decided to include it in the book as an illustrative practical example.

Let us suppose that the Jones family is going to buy a new car. They consider four different vehicles, called car 1, car 2, car 3, and car 4, respectively. During a discussion, they identified that there are five important factors they want to consider when buying a car. These are: cost, safety, design, capacity and warranty. However, some of the criteria can be considered in the context of several different aspects. For example, the cost of a car is not only the purchase price but also the cost of maintenance, inspections, operation and fuel, etc. For this reason, the cost criterion has been divided into the purchase price, fuel cost and the maintenance cost. In the same way, the capacity criterion has been divided into two other sub-criteria: trunk size and the passenger capacity. The Jones family decided that the three other criteria do not have to be divided into more detailed ones. As a result, a four-level model was created (Fig. 1.7) composed of alternatives, criteria, subcriteria and the goal.

---

[9]https://en.wikipedia.org/wiki/Analytic_hierarchy_process_-_car_example

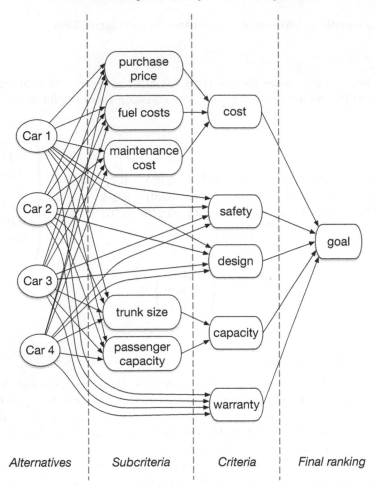

Figure 1.7: Schema of the hierarchy – car selection problem

In the first step, the family compared all four cars with respect to the subcriteria. This results in the five 4×4 matrices containing comparisons of the four considered cars with respect to the purchase prices – $C_{1,1}$, fuel costs – $C_{2,1}$, maintenance costs – $C_{3,1}$, trunk size – $C_{4,1}$ and the passenger capacities – $C_{5,1}$.

$$
C_{1,1} = \begin{pmatrix} 1 & \frac{7}{5} & \frac{4}{9} & \frac{4}{5} \\ \frac{5}{7} & 1 & \frac{6}{7} & \frac{7}{6} \\ \frac{9}{4} & \frac{7}{6} & 1 & \frac{3}{2} \\ \frac{5}{4} & \frac{6}{7} & \frac{2}{3} & 1 \end{pmatrix}, \quad
C_{2,1} = \begin{pmatrix} 1 & \frac{7}{3} & \frac{9}{5} & 2 \\ \frac{3}{7} & 1 & \frac{8}{5} & \frac{8}{5} \\ \frac{5}{9} & \frac{5}{8} & 1 & 2 \\ \frac{1}{2} & \frac{5}{8} & \frac{1}{2} & 1 \end{pmatrix},
$$

$$C_{3,1} = \begin{pmatrix} 1 & \frac{7}{5} & \frac{4}{3} & \frac{5}{9} \\ \frac{5}{7} & 1 & 2 & \frac{6}{5} \\ \frac{3}{4} & \frac{1}{2} & 1 & \frac{3}{2} \\ \frac{9}{5} & \frac{5}{6} & \frac{2}{3} & 1 \end{pmatrix}, \quad C_{4,1} = \begin{pmatrix} 1 & \frac{6}{5} & \frac{2}{3} & \frac{5}{2} \\ \frac{5}{6} & 1 & \frac{5}{9} & \frac{7}{5} \\ \frac{3}{2} & \frac{9}{5} & 1 & 1 \\ \frac{2}{5} & \frac{5}{7} & 1 & 1 \end{pmatrix},$$

$$C_{5,1} = \begin{pmatrix} 1 & 9 & 9 & \frac{3}{8} \\ \frac{1}{9} & 1 & \frac{2}{3} & \frac{1}{9} \\ \frac{1}{9} & \frac{3}{2} & 1 & \frac{1}{9} \\ \frac{8}{3} & 9 & 9 & 1 \end{pmatrix}.$$

When comparing prices (purchase price, fuel cost and maintenance cost), it was assumed that the lower the price, the more favorable it is. The results obtained correspond more or less to the inverse proportionality of the nominal price, i.e., if the purchase price of the first car is \$20000 and the second car is \$30000 then the comparison c_{12} is $3/2$ which means that the first car is $3/2$ more favorable than car 2. The family also discussed criteria which do not have subcriteria. In this way, they prepared three additional PC matrices reflecting their opinions about the cars as to the safety – $C_{2,2}$, design – $C_{3,2}$ and warranty – $C_{5,2}$.

$$C_{2,2} = \begin{pmatrix} 1 & \frac{2}{5} & \frac{1}{9} & \frac{1}{7} \\ \frac{5}{2} & 1 & \frac{1}{9} & \frac{1}{4} \\ 9 & 9 & 1 & 5 \\ 7 & 4 & \frac{1}{5} & 1 \end{pmatrix}, \quad C_{3,2} = \begin{pmatrix} 1 & \frac{1}{9} & \frac{1}{9} & \frac{1}{9} \\ 9 & 1 & 5 & \frac{9}{8} \\ 9 & \frac{1}{5} & 1 & \frac{7}{9} \\ 9 & \frac{8}{9} & \frac{9}{7} & 1 \end{pmatrix},$$

$$C_{5,2} = \begin{pmatrix} 1 & 9 & \frac{4}{3} & \frac{7}{5} \\ \frac{1}{9} & 1 & \frac{1}{9} & \frac{1}{9} \\ \frac{3}{4} & 9 & 1 & \frac{1}{2} \\ \frac{5}{7} & 9 & 2 & 1 \end{pmatrix}.$$

The above PC matrices translate into the priority vectors for the cars with respect to subcriteria:

$$w_{1,1} = \begin{pmatrix} 0.208 \\ 0.226 \\ 0.343 \\ 0.22 \end{pmatrix}, \quad w_{2,1} = \begin{pmatrix} 0.398 \\ 0.24 \\ 0.213 \\ 0.146 \end{pmatrix}, \quad w_{3,1} = \begin{pmatrix} 0.250 \\ 0.279 \\ 0.216 \\ 0.253 \end{pmatrix},$$

$$w_{4,1} = \begin{pmatrix} 0.290 \\ 0.211 \\ 0.314 \\ 0.183 \end{pmatrix}, \quad w_{5,1} = \begin{pmatrix} 0.341 \\ 0.043 \\ 0.052 \\ 0.563 \end{pmatrix},$$

as well as the cars with the criteria without subcriteria (safety, design and warranty):

$$w_{2,2} = \begin{pmatrix} 0.041 \\ 0.072 \\ 0.663 \\ 0.221 \end{pmatrix}, \quad w_{3,2} = \begin{pmatrix} 0.032 \\ 0.481 \\ 0.189 \\ 0.259 \end{pmatrix}, \quad w_{5,2} = \begin{pmatrix} 0.368 \\ 0.034 \\ 0.248 \\ 0.348 \end{pmatrix}.$$

Sub-criteria are evaluated in the next step, separately, divided into those related to the cost and those associated with the capacity. Creating the ranking of the latter is quite easy. Our family decides that the passenger capacity is three times more important than the trunk size which leads to the ranking vector $w_{4,2} = [0.75, 0.25]$. In the case of purchase price, fuel costs and maintenance cost, creating the 3×3 matrix is necessary. Hence,

$$C_{1,2} = \begin{pmatrix} 1 & 7 & 8 \\ \frac{1}{7} & 1 & 3 \\ \frac{1}{8} & \frac{1}{3} & 1 \end{pmatrix}.$$

The ranking vector corresponding to the above matrix is as follows:

$$w_{1,2} = \begin{pmatrix} 0.776 \\ 0.153 \\ 0.07 \end{pmatrix}.$$

Thus, the purchase price criterion received the weight 0.776, fuel cost criterion got 0.153, and maintenance cost got 0.07. At this point, almost all necessary comparisons have been made. However, the weight of all main criteria has not been determined. Therefore, in the last step of the procedure the Jones family compares all the main criteria together. These comparisons form the matrix:

$$C_{1,3} = \begin{pmatrix} 1 & \frac{7}{5} & 5 & \frac{9}{5} & 8 \\ \frac{5}{7} & 1 & \frac{9}{5} & \frac{7}{5} & \frac{5}{4} \\ \frac{1}{5} & \frac{5}{9} & 1 & \frac{3}{7} & \frac{3}{4} \\ \frac{5}{9} & \frac{5}{7} & \frac{7}{3} & 1 & \frac{7}{9} \\ \frac{1}{8} & \frac{4}{5} & \frac{4}{3} & \frac{9}{7} & 1 \end{pmatrix}$$

which leads to the ranking vector:

$$w_{1,3} = \begin{pmatrix} 0.447 \\ 0.193 \\ 0.082 \\ 0.155 \\ 0.12 \end{pmatrix}.$$

In order to calculate the final ranking, all the ranking vectors need to be appropriately merged. Following (1.3) we may write the ranking vector for the first car as:

$$w(a_1) = w_{1,3}(a_{1,3}) \cdot \sum_{k=1}^{3} w_{1,2}(a_{k,2}) \cdot w_{k,1}(a_{1,1}) +$$

$$+ \, w_{1,3}(a_{2,3}) \cdot w_{2,2}(a_{1,1}) + w_{1,3}(a_{3,3}) \cdot w_{3,2}(a_{1,1}) +$$

$$+ \, w_{1,3}(a_{4,3}) \cdot \sum_{k=1}^{2} w_{4,2}(a_{k,2}) \cdot w_{k,1}(a_{1,1}) + w_{1,3}(a_{5,3}) \cdot w_{5,2}(a_{1,1})$$

Providing that

$$\sum_{k=1}^{3} w_{1,2}(a_{k,2}) \cdot w_{k,1}(a_{1,1}) = 0.776 \cdot 0.208 + 0.153 \cdot 0.398 + 0.07 \cdot 0.25 = 0.24$$

and

$$\sum_{k=1}^{2} w_{4,2}(a_{k,2}) \cdot w_{k,2}(a_{1,1}) = 0.75 \cdot 0.29 + 0.25 \cdot 0.341 = 0.302$$

Thus,

$$w(a_1) = 0.447 \cdot 0.24 + 0.193 \cdot 0.041$$
$$+ \, 0.082 \cdot 0.032 + 0.155 \cdot 0.302$$
$$+ \, 0.12 \cdot 0.368 = 0.2101$$

In a similar way, we may compute the weights for a_2, a_3 and a_4, which are $w(a_2) = 0.188$, $w(a_3) = 0.354$ and $w(a_4) = 0.247$ respectively. Since it eventually turned out that the third car (the third alternative) got the highest weight $w(a_3) = 0.354$, it is also the preferred choice of the family.

Chapter 2

PC Matrices

Whenever experts assess two alternatives, options or elements with each other, certain information is generated – the result of a comparison. This information may take different forms and different representations. Sometimes it can be a real positive number, another time it can be one of several values: greater, smaller or equal. It can be a fuzzy number or a numerical range. It can even be a message – this is an unknown value. Each piece of information is the result of a direct comparison of two elements: the i-th and j-th alternatives. Since we assume that experts know the problem being assessed, i.e., when they compare the same pair of alternatives many times all these comparisons produce the same result, the number of comparisons for n alternatives seems to be limited by n^2. Thus, the natural representation for the set of comparisons is the square matrix $C = [c_{ij}]$ in which c_{ij} elements denote the result of comparison of the i-th and j-th alternatives (see: 1.1)

$$
C = \begin{pmatrix}
1 & c_{12} & \cdots & c_{1n} \\
c_{21} & 1 & \cdots & c_{2n} \\
\vdots & \vdots & \ddots & \vdots \\
c_{n1} & c_{n2} & \cdots & 1
\end{pmatrix} .
$$

For this reason, the PC method, as well as AHP itself, is very often referred to in the context of the pairwise comparisons (PC) matrices. Despite their clear algebraic nature, PC matrices are, above all, a neat way of presenting a set of paired comparisons. Thus, the meaning and form of individual c_{ij} are subordinated to the nature of the given approach (type of matrix). Below, several types of PC matrices are presented. We start with the most popular ones. cardinal PC matrices (usually referred to as PC matrices). then we briefly discuss other interesting representations. The list of PC matrix types does not pretend to be complete because new representations of decision data are still emerging (the so-called continuous pairwise comparisons can serve as a perfect example of this statement [166]).

2.1   Cardinal PC matrix

AHP uses multiplicative cardinal matrices, that is, the result of a comparison of two alternatives is interpreted as the ratio of their priorities. For example, if c_{ij} is 5 then it means that the i-th alternative is 5 times more important than the j-th alternative. For the same reason, a diagonal is composed of 1, which obviously means that each alternative compared with itself is equally important. For this type of matrix, we will assume that $c_{ij} \in \mathbb{R}_+^n$ i.e. the result of a direct comparison is a real and positive (greater than 0) number. Although c_{ij} is a result of comparison but not a comparison itself, very often in this book we will refer to c_{ij} as to the comparison between the i-th and j-th alternatives. We hope that this useful abuse of terminology will, on the one hand, improve the readability of the text and, on the other hand, not adversely affect the understanding of the issues discussed. Let us define that kind of matrix formally.

Definition 1. *A cardinal, multiplicative pairwise comparisons matrix (hereinafter referred to as a PC matrix) for n alternatives is said to be the matrix $C = [c_{ij}]$ such that $c_{ij} \in \mathbb{R}_+^n$, $c_{ii} = 1$ for $i, j = 1, \ldots, n$.*

The fact that c_{ij} corresponds to the numerical value of relations between the priorities of two alternatives implies that c_{ji} should express the opposite relationship. That is, if $c_{ij} \approx \frac{importance(a_i)}{importance(a_j)}$, then it is natural to expect that $c_{ji} \approx \frac{importance(a_j)}{importance(a_i)}$, where $importance(a_i)$ represents the internal judgment of an expert regarding the significance of the i-th alternative. The expert does not need to know $importance(a_i)$ or $importance(a_j)$ explicitly (he/she is asked to provide c_{ij}). However, the fact that behind every c_{ij} an appropriate ratio is hidden leads to a natural expectation that $c_{ij} = 1/c_{ji}$. This property is called reciprocity.

Definition 2. *The PC matrix $C = [c_{ij}]$ is said to be reciprocal if for every $i, j = 1, \ldots, n$ it holds that $c_{ij} = 1/c_{ji}$.*

Although in the case of PC matrices, the reciprocity property seems to be natural, non-reciprocal matrices are also considered in the literature [116, 82]. Sometimes, in practice, even comparing the same alternatives to themselves may not always yield 1 (e.g. blind wine tasting) [105]. We had the opportunity to see examples of (cardinal, multiplicative) PC matrices in the section

Examples of (cardinal, multiplicative) PC matrices can be found in (Sec. 1.3).

2.2 Ordinal PC matrix

It is not always possible to determine the ratio between the importance of the i-th and j-th alternatives. Thus, one can at most indicate which alternative is more preferred than the other, but not to what extent. Let us define the preference relation $\prec \subseteq \mathscr{A} \times \mathscr{A}$ where $\mathscr{A} = \{a_1, \ldots, a_n\}$ is the set of alternatives (Sec. 1.2.2). We will say that $a_i \prec a_j$ if a_i is less preferred than a_j (and a_j is more preferred than a_i). It is also useful to allow the existence of ties. This situation means that the expert is preferentially indifferent to the two considered alternatives. In the event of a tie between the i-th and j-th alternatives, we will write $a_i \sim a_j$. The preference relation can also be written in the form of a matrix. In this case, however, the matrix elements must correspond to the type of relationship between the two alternatives.

Definition 3. *An ordinal PC matrix for n alternatives is said to be the matrix* $O = [o_{ij}]$ *such that* $o_{ij} \in \{-1, 0, 1\}$, $o_{ii} = 1$ *for* $i, j = 1, \ldots, n$.

We write that if $a_i \prec a_j$ then $o_{ij} = -1$, $a_j \prec a_i$ then $o_{ij} = 1$, and if $a_i \sim a_j$ then $o_{ij} = 0$. For convenience, we will also use $a_j \succ a_i$ wherever $a_i \prec a_j$, and $a_i \preceq a_j$ if either $a_i \prec a_j$ or $a_i \sim a_j$.

Similarly to the case of a cardinal PC matrix, it is natural to expect that if $o_{ij} = 1$ then also $o_{ji} = -o_{ij} = -1$. Thus, similarly to before, we may define the reciprocity property.

Definition 4. *The ordinal PC matrix* $O = [o_{ij}]$ *is said to be reciprocal if for every* $i, j = 1, \ldots, n$ *it holds that* $o_{ij} = -o_{ji}$.

Example 1. *Let us consider the following example. A man wants to choose the color he will paint a house. The seller offered him four "popular" colors:* $a_1, \ldots, a_4$. *A can of each paint costs the same, so the price does not affect the decision. As a result of comparing paints in pairs, the following ordinal PC matrix has been formed:*

$$C = \begin{pmatrix} 0 & 1 & 1 & -1 \\ -1 & 0 & 1 & -1 \\ -1 & -1 & 0 & -1 \\ 1 & 1 & 1 & 0 \end{pmatrix}.$$

Since, in direct comparisons, the fourth color wins against any other color, i.e., $a_4 \succ a_i$ where $i = 1, 2, 3$, then a_4 is the winner of the ranking.

Every cardinal PC matrix can also be considered as an ordinal PC matrix. Let $C = [c_{ij}]$ be a cardinal pc matrix, then $O = [o_{ij}]$ such that

$o_{ij} = \text{sign}(\log c_{ij})$ is a corresponding ordinal PC matrix, where

$$\text{sign}(x) = \begin{cases} 1 & \text{if } x > 0 \\ 0 & \text{if } x = 0 \\ -1 & \text{if } x < 0 \end{cases}.$$

Hence, wherever a_i is more important than a_j i.e. if $c_{ij} > 1$ then also $o_{ij} = 1$ etc. For example, the ordinal counterpart of $C_{1,2}$ (1.4) from (Sec. 1.3.1) is

$$O = \begin{pmatrix} 0 & 1 & 1 & 1 & 1 & 1 & 1 & 1 \\ -1 & 0 & 1 & 1 & 1 & 1 & 1 & -1 \\ -1 & -1 & 0 & -1 & 1 & 1 & 1 & -1 \\ -1 & -1 & 1 & 0 & 1 & 1 & 1 & -1 \\ -1 & -1 & -1 & -1 & 0 & -1 & -1 & -1 \\ -1 & -1 & -1 & -1 & 1 & 0 & -1 & -1 \\ -1 & -1 & -1 & -1 & 1 & 1 & 0 & -1 \\ -1 & 1 & 1 & 1 & 1 & 1 & 1 & 0 \end{pmatrix}.$$

Sometimes it is worth considering a matrix without ties. As comparing an alternative with itself must lead to a tie, we are not able to put any specific numerical value on the diagonal. Thus, in such a case we will put a dash on the diagonal. This means that this type of comparison is undefined.

2.3   Incomplete PC matrix

In practice, gathering all the results of pairwise comparisons for n alternatives can be difficult, time-consuming and even impossible. In such a case, the value of such comparisons is omitted and the ranking is calculated only on the basis of existing comparisons. In order to mark missing comparisons we will use the symbol ?. Formally, the incomplete PC matrix can be defined as follows.

Definition 5. *A cardinal (multiplicative) incomplete pairwise comparisons matrix for n alternatives is said to be the matrix $C = [c_{ij}]$ such that $c_{ij} \in \mathbb{R}_+ \cup \{?\}$. C is said to be reciprocal if for every $c_{ij} \in \mathbb{R}_+$ it holds that $c_{ji} = 1/c_{ij}$, and if $c_{ij} =?$ then also $c_{ji} =?$.*

In a similar way, an ordinal incomplete PC matrix can be defined.

Definition 6. *An ordinal incomplete pairwise comparisons matrix for n alternatives is said to be the matrix $O = [o_{ij}]$ such that $o_{ij} \in \{-1, 0, 1, ?\}$. O is said to be reciprocal if for every $o_{ij} \in \{-1, 0, 1\}$ it holds that $o_{ij} = -o_{ji}$, and if $o_{ij} =?$ then also $o_{ji} =?$.*

The graph of an incomplete PC matrix does not contain edges corresponding to the missing comparisons. For example, for an incomplete ordinal PC matrix

$$C = \begin{pmatrix} 0 & 1 & ? & 1 & 0 \\ -1 & 0 & ? & ? & 1 \\ ? & ? & 0 & 1 & ? \\ -1 & ? & -1 & 0 & 1 \\ 0 & -1 & ? & -1 & 0 \end{pmatrix},$$

the corresponding graph T_C looks as follows (Fig. 2.1).

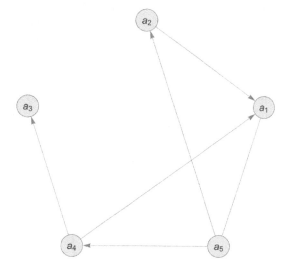

Figure 2.1: Graph of the incomplete ordinal PC matrix C

To be able to calculate the ranking of n alternatives, an indirect relationship must exist between each two. In terms of the graph, for every two alternatives (vertices) there must be a path between them. Such a graph in which this condition is met is said to be connected.

Definition 7. *The irreducible matrix is one that cannot be decomposed into the form:*

$$C = \begin{pmatrix} Q_1 & ? \\ Q_2 & Q_3 \end{pmatrix}$$

where Q_1, Q_3 are square matrices, and ? denotes a matrix composed of undefined values[1].

The matrix C corresponding to the connected graph G_C is irreducible [76, 146].

---

[1]In the classical definition of a graph matrix, the missing edges are represented by 0 [146]. Due to the need to represent ties in the ordinal PC matrices, we decided to use a question mark for this purpose.

2.4   PC matrix as a graph

It is often convenient to consider a set of pairwise comparisons as a graph (both directed and undirected). In this approach, vertices correspond to alternatives, whereas edges represent comparisons. The labels of edges may denote the values of comparisons. Let us introduce the definition of a graph (first directed then undirected) of the given PC matrix C.

Definition 8. *The directed graph $T_C = (A, E, L)$ is said to be a (directed) graph (when it is a complete graph also called a tournament, or a t-graph) of the PC matrix $C = [c_{ij}]$ if $A = \{a_1, \ldots, a_n\}$ is a set of vertices, $E \subseteq \{(a_i, a_j) \in A \times A : i \neq j \text{ and } c_{ij} \text{ is defined}\}$ is a set of edges, and $L : A \times A \to \mathbb{R}$ is the labeling function such that $L(a_i, a_j) = c_{ij}$.*

In the above definition, we used the same symbol A to denote the set of vertices in the graph and the set of alternatives. Although these are different objects, they will always be uniquely related. Therefore, to reduce the number of symbols used, we have decided on this "convenient" abuse. Let us consider

$$C = \begin{pmatrix} 1 & 9 & \frac{5}{8} & 7 & 9 \\ \frac{1}{9} & 1 & \frac{1}{9} & \frac{1}{6} & \frac{6}{7} \\ \frac{8}{5} & 9 & 1 & 6 & 9 \\ \frac{1}{7} & 6 & \frac{1}{6} & 1 & \frac{9}{4} \\ \frac{1}{9} & \frac{7}{6} & \frac{1}{9} & \frac{4}{9} & 1 \end{pmatrix}.$$

According to Def. 8, the graph of C is given as T_C (Fig. 2.2).

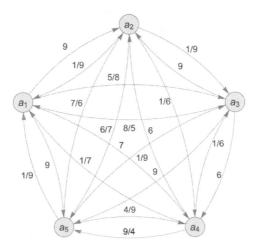

Figure 2.2: T_C – graph of the matrix C

Even though T_C contains all the information provided by C, the graphical representation of the graph may not be easy to understand. The reason is the quite large number of edges and their labels. One way to reduce the number of edges is to consider only the upper triangular matrix of C. Thanks to the reciprocity condition, the upper triangular matrix of C contains all the same information as the whole PC matrix. Let

$$U(C) = \begin{cases} c_{ij} & if\ i < j \\ ? & otherwise \end{cases},$$

where the symbol ? means that the entry is undefined. Then the reciprocal matrix C can also be unambiguously represented by the graph $T_{U(C)}$ (Fig. 2.3).

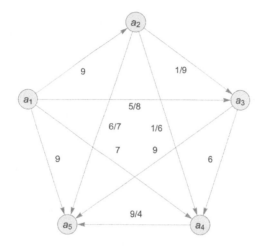

Figure 2.3: $T_{U(C)}$ – graph of the upper triangle matrix of C

Since both graphs T_C and $T_{U(C)}$ contain the same information (at least as long as C is reciprocal), $T_{U(L)}$ will often be used.

If the direction of comparisons is not important to us, but the fact that two alternatives are compared is, we may also use an undirected graph instead of a directed one.

Definition 9. *The undirected graph $P_C = (A, E)$ is said to be a (undirected) graph of the PC matrix $C = [c_{ij}]$ if $A = \{a_1, \dots, a_n\}$ is a set of vertices, $E \subseteq \{\{a_i, a_j\} \in 2^A\ :\ i \neq j\ and\ c_{ij}\ is\ defined\}$ is a set of edges.*

It is worth noting that P_C is like a "reduced" version of $T_{U(C)}$. Indeed, providing that C is reciprocal, we can get P_C from $T_{U(C)}$ by "removing" labels and "arrows" on the edges.

Another useful concept from graph theory is the path between two vertices. Formally,

Definition 10. *An ordered sequence of distinct vertices $p = a_{i_1}, a_{i_2}, \dots, a_{i_m}$ such that $\{a_{i_1}, a_{i_2}, \dots, a_{i_m}\} \in V$ is said to be a path between a_{i_1} and a_{i_m} with*

the length $m-1$ in $T_C = (V, E, L)$ if $\{a_{i_1}, a_{i_2}\}, \{a_{i_2}, a_{i_3}\}, \ldots, \{a_{i_{m-1}}, a_{i_m}\} \in E$.
A path is called elementary if its elements do not repeat.

From the point of view of the above definition, the edge direction is not important. There is a transition between one vertex and the other if there is any mutual comparison, regardless of its outcome. In fact, in the case of reciprocal PC matrices, it is not important whether we compare a_i with a_j or, vice versa, first a_j then a_i. A special kind of path is a cycle.

Definition 11. *A path between $a_{i_1}, a_{i_k} \in A$ for the graph $P_C = (A, E)$ such that $a_{i_1} = a_{i_k}$ is called a cycle. A cycle is called elementary if it is an elementary path.*

Both of the above concepts can be interpreted in the context of PC matrices. As will be shown later, paths and cycles can be used to determine the degree of inconsistency in the pairwise comparison matrix. For example, a path between two vertices $a_{i_1}, a_{i_k} \in A$ in the graph $P_C = (A, E)$ forms a kind of indirect relationship between the corresponding alternatives a_{i_1} and a_{i_k}. Indeed, it is natural to expect that for a path $a_{i_1}, \ldots, a_{i_k}$ holds $w(a_{i_1}) \approx w(a_{i_k}) \cdot \sum_{r=1}^{k-1} c_{i_r, i_{r+1}}$. In particular, when C is consistent then $w(a_{i_1}) = w(a_{i_k}) \cdot \sum_{r=1}^{k-1} c_{i_r, i_{r+1}}$ (Theorem 4).

One may observe that the minimum number of comparisons necessary to determine the ranking of n alternatives is $n-1$. Indeed, when we compare alternative a_1 with a_2, then a_2 with a_3, and so on up to the comparison a_{n-1} with a_n, then we obtain the set of $n-1$ comparisons $Q = \{c_{1,2}, c_{2,3}, \ldots, c_{n-1,n}\}$ that can be used to compute the ranking. To calculate it, we may just assume $w(a_1) = 1$ then assign $w(a_2) = c_{1,2} \cdot w(a_1)$, $w(a_3) = c_{2,3} \cdot w(a_2)$ and so on. The alternatives (vertices) $A = \{a_1, \ldots, a_n\}$ together with the edges $\{a_1, a_2\}, \{a_2, a_3\}, \ldots, \{a_{n-1}, a_n\}$ form the sub-graph of P_C called a spanning tree of P_C.

Definition 12. *A spanning tree of the graph $P_C = (A, E)$ is said to be the graph $S = (A, Q)$ such that $Q \subseteq E$, $|Q| = |A| - 1$ and for every two vertices a_i and a_j there is a path in S.*

There are many spanning trees for the given graph[2]. Interestingly, comparisons determined by any of the spanning trees can be used to determine unambiguous rankings of all the considered alternatives.

Not only cardinal matrices can be represented in the form of a graph. Representations in the form of a graph also have ordinal matrices. Since, in the case of an ordinal PC matrix, the results of comparisons are limited to $\{-1, 0, 1\}$, then we can opt out of using labels and use properly directed (or undirected) edges instead. Following [113], let us introduce the notion of a graph of C, where C is an ordinal PC matrix.

---

[2]Following Cayley's formula, there are n^{n-2} spanning trees for a complete undirected graph with n vertices [37].

Definition 13. *A graph of the ordinal $n \times n$ PC matrix C (when it is complete graph also called a generalized tournament, or briefly a gt-graph) is a triple $G_C = (A, E_u, E_d)$ where $A = \{a_1, \ldots, a_n\}$ is a set of vertices (alternatives), $E_u \subset 2^A$ is a set of unordered pairs called undirected edges $E_u \subseteq \{\{a_i, a_j\} : a_i, a_j \in A, c_{ij} = 0 \text{ and } i \neq j\}$, and $E_d \subseteq V^2$ is a set of ordered pairs called directed edges $E_d \subseteq \{(a_i, a_j) : a_i, a_j \in A, c_{ji} = 1 \text{ and } i \neq j\}$. G_C does not contain doubled edges, i.e. for every $i, j = 1, \ldots, n$: if $(a_i, a_j) \in E_d$ then $\{a_i, a_j\} \notin E_u$, reversely if $\{a_i, a_j\} \in E_u$ then $(a_i, a_j) \notin E_d$.*

For example, a graph for an ordinal PC matrix:

$$C = \begin{pmatrix} 0 & 1 & 0 & 1 & 0 \\ -1 & 0 & 1 & 1 & 1 \\ 0 & -1 & 0 & 1 & -1 \\ -1 & -1 & -1 & 0 & 1 \\ 0 & -1 & 1 & -1 & 0 \end{pmatrix}$$

is shown in (Fig. 2.4).

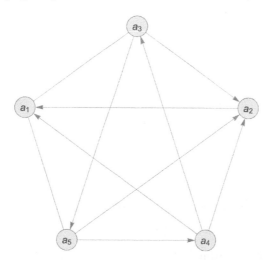

Figure 2.4: Graph of the ordinal PC matrix C

Finally, following [119] let us define families of paths and cycles introduced by the given (incomplete) PC matrix.

Definition 14. *Let $T_C = (V, E, L)$ be a graph of C. The set of all paths between a_i and a_j in T_C is denoted as $\mathcal{P}_{C,i,j} \overset{df}{=} \{p = a_{i_1}, a_{i_2}, \ldots, a_{i_m} \text{ is a path between } a_{i_1} \text{ and } a_{i_m} \text{ in } T_C\}$. Similarly, the set of all cycles longer than q in T_C is denoted as $\mathcal{S}_{C,q} \overset{df}{=} \{s = a_{i_1}, a_{i_2}, \ldots, a_{i_m} \text{ is a cycle of } C \text{ where } m > q\}$.*

2.5   Graph as a matrix

As shown above, a PC matrix can be presented in the form of a graph. However, a graph as such can also be presented in the form of a matrix and, thanks to such a representation, it is possible to conclude its properties. Indeed, when discussing ranking procedures for incomplete matrices we use the notions defined below.

One of the basic characteristics associated with the vertex in the graph is the number of adjacent edges. We can distinguish incoming, outgoing and undirected edges, or count them all together regardless of their direction. Hence, with every vertex $a \in A$ (we will continue to denote vertices in a graph in the same manner as alternatives in a ranking model) we can associate a series of numbers (degrees) describing its connections.

The simplest case is an undirected graph.

Definition 15. *The degree of $a_i \in V$, where $P_C = (A, E)$ is a graph of the (cardinal) PC matrix C, denoted $\deg(a_i, G_C)$ correspondingly are defined as follows:*

- $\deg(a_i) = |\{a_j \mid \exists\{a_j, a_i\} \in E\}|$

For directed graphs, the matter becomes complicated.

Definition 16. *The in-degree, out-degree and degree of $a_i \in V$, where $T_C = (A, E, L)$ is a graph of the (cardinal) PC matrix C, denoted $\deg_{in}(a_i, T_C), \deg_{out}(a_i, T_C)$ and $\deg(a_i, T_C)$ correspondingly are defined as follows:*

- $\deg_{in}(a_i) = |\{a_j \mid \exists(a_j, a_i) \in E\}|,$

- $deg_{out}(a_i) = |\{a_j \mid \exists(a_i, a_j) \in E\}|$ *and*

- $deg(a_i) = \deg_{in}(a_i) + deg_{out}(a_i).$

In the case of an ordinal PC matrix and its graph representation, the situation becomes even more complicated than before. This is because, in addition to directed edges, undirected edges can also be found in the graph, then also the number of undirected edges adjacent to the given vertex has to be taken into account.

Definition 17. *The in-degree, out-degree, un-degree and degree of the vertex $a_i \in V$, where $G_C = (A, E_u, E_d)$ is a graph of the (ordinal) PC matrix C, denoted $\deg_{in}(a_i), \deg_{out}(a_i), \deg_{un}(a_i)$ and $\deg(a_i)$ correspondingly are defined as follows:*

- $\deg_{in}(a_i) = |\{a_j \mid \exists(a_j, a_i) \in E_d\}|,$

- $deg_{out}(a_i) = |\{a_j \mid \exists(a_i, a_j) \in E_d\}|,$

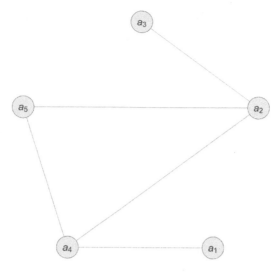

Figure 2.5: G_C graph

- $deg_{un}(a_i) = |\{a_j \mid \exists e \in E_u : a_j \in e\}|$,
- $deg(a_i) = deg_{in}(a_i) + deg_{out}(a_i) + deg_{un}(a_i)$.

Despite the similarities the above definitions differ slightly. For this, depending on the context (type of graph) the reader will have to use the right one. In the work we adopted the principle to use P_C for denoting undirected graphs, T_C for denoting directed (in particular for a tournament) graphs and G_C for denoting mixed graphs (graphs containing directed and undirected edges).

Based on the concept of the apex degree in the graph, we can define the concept of a degree matrix.

Definition 18. *The degree matrix of* $P_C = (A, E)$ *is the matrix* $D(P_C) = [d_{ij}]$ *such that*

$$d_{ij} = \begin{cases} deg(a_i) & if\ i = j \\ 0 & otherwise \end{cases}.$$

For example, for some graph P_C as shown in (Fig. 2.5) the degree matrix looks as follows:

$$D(G_C) = \begin{pmatrix} 1 & 0 & 0 & 0 & 0 \\ 0 & 3 & 0 & 0 & 0 \\ 0 & 0 & 1 & 0 & 0 \\ 0 & 0 & 0 & 3 & 0 \\ 0 & 0 & 0 & 0 & 2 \end{pmatrix}.$$

The neighborhood relation between the vertices in the graph is represented by an adjacency matrix $N = [n_{ij}]$, where $n_{ij} \in \{0, 1\}$.

Definition 19. *The adjacency matrix of $P_C = (A, E)$ is the matrix $N(P_C) = [n_{ij}]$ such that*

$$n_{ij} = \begin{cases} 0 & if\ i = j\ or\ \{a_i, a_j\} \in E \\ 1 & if\ i \neq j\ and\ \{a_i, a_j\} \notin E \end{cases}.$$

It is easy to notice that, as long as C is reciprocal, the adjacency matrix $N(P_C)$ is symmetric. Its entries denote the existence of a comparison between alternatives. That is, if $n_{ij} = 1$ then there are comparisons between the i-th and j-th alternatives.

The combination of the degree matrix and the adjacency matrix is the Laplacian matrix:

$$L(P_C) = D(P_C) - N(P_C).$$

The Laplacian matrix, sometimes called the admittance matrix or Kirchhoff matrix, has many intriguing properties [128]. In particular, one may show that it is singular, and when the graph G_C is strongly connected then all but the smallest one of its eigenvalues are real and greater than zero. The smallest principal eigenvalue λ_0 equals 0 etc. More information about the Laplacian matrix can be found in [128, 133].

2.6   Additive PC matrix

In most cases, the elements of the pairwise comparison matrix represent the ratio of the preferences of one alternative to the other. Hence, it is natural to expect that $c_{ij} \approx w(a_i)/w(a_j)$. However, this is not the only way to represent the relationship of preferences between alternatives. Another one is the difference in preferential values. In such a case, $c_{ij} \approx w(a_i) - w(a_j)$. This, of course, implies that $c_{ji} \approx w(a_j) - w(a_i)$ and $c_{ii} = 0$. Formally, an additive PC matrix can be defined as below.

Definition 20. *An additive PC matrix for n alternatives is said to be the matrix $C = [c_{ij}]$ such that $c_{ij} \in \mathbb{R}$, $c_{ii} = 0$ for $i, j = 1, \ldots, n$. C is said to be reciprocal if $c_{ij} = -c_{ji}$.*

Every multiplicative PC matrix can be transformed into an additive one and reversely. That is, if $Q = [q_{ij}]$ is an additive PC matrix, then $R = [e^{q_{ij}}]$ is its multiplicative equivalent. Reversely, if $R = [r_{ij}]$ is a multiplicative PC matrix then $Q = [\ln r_{ij}]$ is its additive representation [13, 14].

2.7 Fuzzy PC matrix

The idea of a fuzzy PC matrix is a response to the need for a representation of uncertain preferential information. Hence, in the case of fuzzy comparisons, their result is a fuzzy number. To define a fuzzy number, first let us introduce the notion of a fuzzy set [153].

Definition 21. *The fuzzy set* Q *for the given space* $X, X \neq \emptyset$ *is the set of pairs*

$$Q = \{(x, \mu_Q(x)) \text{ such that } x \in X\},$$

where

$$\mu_Q : X \to [0, 1]$$

is the membership function of the fuzzy set Q.

The μ_Q determines the membership degree of x. Usually, it is convenient to distinguish three cases:

- full membership, when $\mu_Q(x) = 1$, i.e. $x \in Q$,

- partial membership, when $0 < \mu_Q(x) < 1$, and

- lack of membership, when $\mu_Q(x) = 0$, i.e. $x \notin Q$.

Fuzzy sets defined over $\mathbb{R}$ and that meet three additional conditions (normality, convexity and continuity) are called fuzzy numbers.

Definition 22. *The fuzzy set* Q *such that* $X = \mathbb{R}$, *i.e.* $\mu_Q : \mathbb{R} \to [0, 1]$ *where*

- $\sup_{x \in \mathbb{R}} \mu_Q(x) = 1$ *(normality),*

- $\mu_Q (\lambda x_1 + (1 - \lambda)x_2) \geq \min\{\mu_Q(x_1), \mu_Q(x_2)\}$ *(convexity),*

- μ_Q *is at least segmentally continuous (continuity)*

is called a fuzzy number.

In addition to a fairly simple interpretation, the formalism of fuzzy sets (fuzzy numbers) provides a set of operations such as union, intersection, difference or inclusion and others [153]. On fuzzy numbers, there are also defined arithmetic operations. For example, providing that Q_1 and Q_2 are fuzzy numbers, addition, multiplication and inversion can be defined as follows:

- B is the sum of Q_1 and Q_2, i.e. $B = Q_1 \oplus Q_2$
  if $\mu_B(x) \overset{df}{=} \sup_{x_1, x_2, y = x_1 + x_2} \min \{\mu_{Q_1}(x_1), \mu_{Q_2}(x_2)\}$,

- B is the product of Q_1 and Q_2, i.e. $B = Q_1 \odot Q_2$
  if $\mu_B(x) \overset{df}{=} \sup_{x_1, x_2, y = x_1 \cdot x_2} \min \{\mu_{Q_1}(x_1), \mu_{Q_2}(x_2)\}$,

- B is the inverse of Q i.e. $B = Q^{-1}$ if $\mu_B(x) \overset{df}{=} \mu_Q(x^{-1})$.

Example 2. *Let $Q_1 = \{(2, 0.7), (3, 1), (4, 0.6)\}$ and $Q_2 = \{(3, 0.8), (4, 1), (6, 0.5)\}$ be two fuzzy numbers. Then, according to the above definitions, their sum is given as*

$$
\begin{aligned}
Q_1 \oplus Q_2 = \{ &(5, \min\{0.7, 0.8\}),\ (6, \max\{\min\{0.7, 1\}, \min\{1, 0.8\}\}), \\
&(7, \max\{\min\{1, 1\}, \min\{0.6, 0.8\}\}), \\
&(8, \max\{\min\{0.7, 0.5\}, \min\{0.6, 1\}\}), \\
&(9, \min\{1, 0.5\}),\ (10, \min\{0.6, 0.5\})\},
\end{aligned}
$$

hence

$$
Q_1 \oplus Q_2 = \{(5, 0.7), (6, 0.8), (7, 0.1), (8, 0.6), (9, 0.5), (10, 0.5)\}
$$

It is easy to see that the key element needed to determine a fuzzy number is its membership function μ_Q. Examples of fuzzy membership functions are: a triangular function,

$$
\mu^{(1)}_{l,m,u}(x) = \max\left(\min\left(\frac{x-l}{m-l}, \frac{u-x}{u-m}\right), 0\right),
$$

a trapezoidal function

$$
\mu^{(2)}_{l,m_1,m_2,u}(x) = \max\left(\min\left(\frac{x-l}{m_1-l}, 1, \frac{u-x}{u-m_2}\right), 0\right),
$$

a Gaussian function

$$
\mu^{(3)}_{c,\sigma}(x) = e^{-\frac{1}{2}\left(\frac{x-c}{\sigma}\right)^2},
$$

and a bell-shaped function

$$
\mu^{(4)}_{a,b,c}(x) = \frac{1}{1 + \left|\frac{x-c}{a}\right|^{2b}}.
$$

See (Fig. 2.6).

The most popular one, the triangular function $\mu^{(1)}_{l,m,u}(x)$, underlies the so-called triangular fuzzy numbers . Let us denote a triangular fuzzy number with the membership function $\mu^{(1)}_{l,m,u}(x)$ as triple (l, m, u). Three parameters l, m and u mean successively the smallest possible, the most likely, and the greatest possible values that the uncertain quantity represented by[3] the fuzzy set may take. For these numbers, the arithmetic operations can be defined as the operations on their parameters [194, 123, 27],

$$
(l_1, m_1, u_1) \oplus (l_2, m_2, u_2) \overset{df}{=} (l_1 + l_2, m_1 + m_2, u_1 + u_2), \qquad (2.1)
$$

---

[3] Precisely, for the triangular fuzzy number Q determined by (a, b, c) there exists $p_1, p_2 > 0$ such that $(a + \epsilon, p_1), (b, 1), (c - \epsilon, p_2) \in Q$.

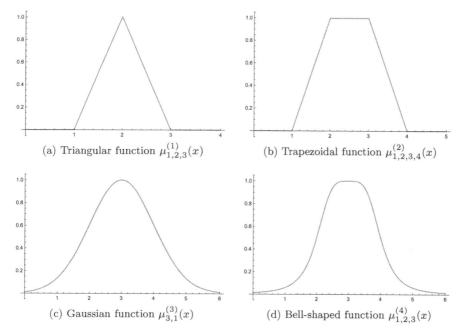

(a) Triangular function $\mu_{1,2,3}^{(1)}(x)$

(b) Trapezoidal function $\mu_{1,2,3,4}^{(2)}(x)$

(c) Gaussian function $\mu_{3,1}^{(3)}(x)$

(d) Bell-shaped function $\mu_{1,2,3}^{(4)}(x)$

Figure 2.6: Fuzzy membership functions

$$(l_1, m_1, u_1) \ominus (l_2, m_2, u_2) \overset{df}{=} (l_1 - l_2, m_1 - m_2, u_1 - u_2), \qquad (2.2)$$

$$(l_1, m_1, u_1) \odot (l_2, m_2, u_2) \overset{df}{=} (l_1 \cdot l_2, m_1 \cdot m_2, u_1 \cdot u_2), \qquad (2.3)$$

$$(l, m, u)^{-1} \overset{df}{=} (l^{-1}, m^{-1}, u^{-1}), \qquad (2.4)$$

$$\ln(l, m, u) \overset{df}{=} (\ln l, \ln m, \ln u), \qquad (2.5)$$

$$e^{(l,m,u)} \overset{df}{=} (e^l, e^m, e^u). \qquad (2.6)$$

It is worth noting that multiplication and inverse are defined as approximations. This means, for example, that there is a slight difference between $\mu_{l,m,u}^{(1)}(x^{-1})$ and $\mu_{l^{-1},m^{-1},u^{-1}}^{(1)}(x)$ (Fig. 2.7). In practice, however, this difference is neglected.

Definition 23. *A fuzzy PC matrix for n alternatives is the matrix $C = [c_{ij}]$ such that c_{ij} is a fuzzy number. C is said to be reciprocal if $c_{ij} = c_{ji}^{-1}$.*

An example of a triangular fuzzy PC matrix may look as follows [27]:

$$C = \begin{pmatrix} (1,1,1) & (\frac{1}{2},2,3) & (1,1,2) \\ (\frac{1}{3},\frac{1}{2},2) & (1,1,1) & (\frac{1}{3},2,4) \\ (\frac{1}{2},1,1) & (\frac{1}{4},\frac{1}{2},3) & (1,1,1) \end{pmatrix},$$

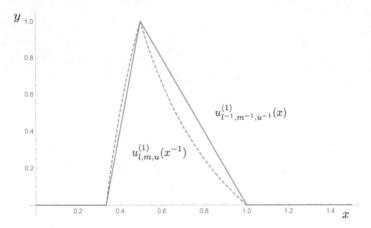

Figure 2.7: The difference between triangular fuzzy number with the inverse argument and its inversion, for $l = 1, m = 2$, and $u = 3$

where $(1, 1, 1) = \{(1, 1)\} \cup \mathbb{R}_+ \backslash \{1\} \times \{0\}$.

More about the fuzzy PC matrices one can read in the work of *Ramik, Gavalec,* [70, 147], *Yuen* [197, 199] or *Kahraman* [93].

Chapter 3

Prioritization Methods

Whenever we needed the concept of ranking in Chapter 1, we created it just like that. We conjured it out of thin air with a click of the fingers (or, if you like, a mysterious prioritization method), each time explaining that it resulted from a specific PC matrix. In this chapter, we will try to make up for these shortcomings. In particular, we will explain the basics of the two fundamental mean-based prioritization methods for quantitative PC matrices [116]. These are the Eigenvalue Method, abbreviated as EVM, and the Geometric Mean Method, shortened to GMM. In addition to these two methods, there are a number of ways to calculate the priority vector based on optimization methods, column averaging or cosine calculation. Some of them will also be presented in this chapter.

3.1   Eigenvalue method

3.1.1   Idea of the method

To explain the idea that led to the creation of EVM, let us recall that, according to the construction of the PC matrix $C = [c_{ij}]$, every entry c_{ij} should express the relative value of the i-th alternative in comparison with the j-th alternative. This is at least the principle according to which the experts assign a certain value to c_{ij}. On the other hand, we want to construct a mechanism (perhaps a function) which gets all such comparisons gathered in C and produces one priority vector at the end:

$$w = \begin{pmatrix} w(a_1) \\ w(a_2) \\ \vdots \\ w(a_n) \end{pmatrix}, \tag{3.1}$$

where every $w(a_i)$ corresponds to the priority value of the i-th alternative. Since, according to the expert judgment, c_{ij} corresponds to the ratio between the importance of the i-th and j-th alternatives, it is natural to expect that the ratio composed of priorities of the i-th and j-th alternatives equals c_{ij}, i.e.,

$$\frac{w(a_i)}{w(a_j)} = c_{ij}.$$

This leads immediately to the observation that the priority of the i-th alternative can be expressed in terms of the priority of the j-th alternative multiplied by the factor c_{ij},

$$w(a_i) = c_{ij}w(a_j). \tag{3.2}$$

How can we expect reality to be different from our expectations? In practice, it is very often not possible to find such a w that the above condition would be met. If this is not possible, then maybe we can expect that at least $w(a_i)$ is approximately the same as $c_{ij}w(a_j)$, i.e.,

$$w(a_i) \approx c_{ij}w(a_j). \tag{3.3}$$

As it turns out, this condition can be met in all circumstances. Thus, let us suppose that the above equation (3.3) applies to every $i, j = 1, \ldots, n$. In particular, as an approximation of $w(a_i)$ we can take $c_{i1}w(a_1), c_{i2}w(a_2)$, and so on, up to $c_{in}w(a_n)$. Thus, instead of (3.2), which might be difficult to meet, we can request from the priority function w that $w(a_i)$ be the mean of all its approximations. Hence,

$$w(a_i) = \frac{1}{n} \sum_{j=1}^{n} c_{ij}w(a_j), \tag{3.4}$$

i.e.,

$$n \cdot w(a_i) = c_{i1}w(a_1) + c_{i2}w(a_2) + \ldots + c_{in}w(a_n).$$

Of course, we may expect that the above observation holds for every $i = 1, \ldots, n$. This translates to the following linear equation system:

$$
\begin{array}{rcl}
c_{1,1}w(a_1) + c_{1,2}w(a_2) + \ldots\ldots\ldots + c_{1,n}w(a_n) &=& n \cdot w(a_1) \\
c_{2,1}w(a_1) + c_{2,2}w(a_2) + \ldots + c_{2,n}w(a_n) &=& n \cdot w(a_2) \\
\ldots\ldots\ldots\ldots\ldots\ldots\ldots\ldots\ldots & & \\
c_{n,1}w(a_n) + c_{n,2}w(a_2) + \ldots + c_{n,n}w(a_n) &=& n \cdot w(a_n)
\end{array}
$$

Keeping in mind that $c_{i,i} = 1$ for $i = 1, \ldots, n$ we have:

$$
\begin{array}{rcl}
w(a_1) + c_{1,2}w(a_2) + \ldots\ldots\ldots + c_{1,n}w(a_n) &=& n \cdot w(a_1) \\
c_{2,1}w(a_1) + w(a_2) + \ldots + c_{2,n}w(a_n) &=& n \cdot w(a_2) \\
\ldots\ldots\ldots\ldots\ldots\ldots\ldots\ldots\ldots & & \\
c_{n,1}w(a_n) + c_{n,2}w(a_2) + \ldots + w(a_n) &=& n \cdot w(a_n)
\end{array}
$$

The above can be written in the matrix form

$$Cw = nw, \tag{3.5}$$

where

$$
C = \begin{pmatrix}
1 & c_{12} & \cdots & c_{1n} \\
c_{21} & 1 & \cdots & c_{2n} \\
\vdots & \vdots & \ddots & \vdots \\
c_{n1} & c_{n2} & \cdots & 1
\end{pmatrix}
$$

and w is the priority vector as shown in (3.1). Is it possible to solve the equation (3.5) for every possible C? The answer is no. Fortunately, a very similar equation to (3.5) already has a solution. This equation is

$$Cw = \lambda w, \tag{3.6}$$

where λ is some specially selected number called an eigenvalue [146]. In fact, we are not interested in just any eigenvalue (in general λ can also be a complex number) but in the real one. How do we know that such a λ exists? A positive answer is provided by the Perron theorem[1] [67] which says: any positive matrix C has a real and positive eigenvalue which exceeds the moduli of all other eigenvalues[2]. As the PC matrix $C = [c_{ij}]$ is real, i.e., all its entries $c_{ij} \in \mathbb{R}$, and positive, i.e., all its entries $c_{ij} > 0$, then there also exists a real and positive

---

[1]Sometimes in the literature it is referred as the Perron-Frobenius theorem. In fact, there are two separate theorems, where the Frobenius theorem deals with the non-negative matrices, while *Perron* deals with a special case of non-negative matrices, i.e., positive matrices [67, v. II, p. 53].

[2]The Perron and Frobenius theorems refer to the eigenvalues as to the roots of $\det(x \cdot Id - C)$ – a characteristic polynomial of C [67].

λ which meets the equation (3.6). Since the Perron theorem also guarantees that it is the largest eigenvalue, we will call it the principal eigenvalue of C and denote λ_{max}[3]. The vector w that accompanies the principal eigenvalue in (3.6) is called a principal eigenvector. For this reason, if necessary, we will denote it as w_{max}. The Frobenius theorem guarantees that this vector is also positive. There are many numerical methods allowing the equation (3.6) to be solved. A valuable reference is the book *Numerical Mathematics* by *Quarteroni, Sacco and Saleri* [146].

For this reason, we will accept that for a PC matrix C the principal eigenvalue and the principal eigenvector are known (can be easily calculated). Thus, we may write that:

$$Cw_{max} = \lambda_{max}w_{max}. \tag{3.7}$$

However, as we remember that the previous consideration led us to the equation (3.5), the question arises as to whether w_{max} is suitable for a priority vector for C. Let us write down (3.7) in the form of the linear equation system:

$$
\begin{aligned}
w_{max}(a_1) + c_{1,2}w_{max}(a_2) + \ldots\ldots\ldots + c_{1,n}w_{max}(a_n) &= \lambda_{max} \cdot w_{max}(a_1) \\
c_{2,1}w_{max}(a_1) + w_{max}(a_2) + \ldots + c_{2,n}w_{max}(a_n) &= \lambda_{max} \cdot w_{max}(a_2) \\
\ldots\ldots\ldots\ldots\ldots\ldots\ldots\ldots\ldots\ldots\ldots\ldots \\
c_{n,1}w_{max}(a_n) + c_{n,2}w_{max}(a_2) + \ldots + w_{max}(a_n) &= \lambda_{max} \cdot w_{max}(a_n)
\end{aligned}
$$

Thus, the priority of the i-th alternative is given as:

$$w_{max}(a_i) = \frac{1}{\lambda_{max}} \sum_{j=1}^{n} c_{ij}w_{max}(a_j). \tag{3.8}$$

The above equation is very similar to (3.4). The only difference between (3.8) and (3.4) is that the sum component is multiplied by a different coefficient. In fact (3.8) we can even write that:

$$w_{max}(a_i) = \gamma\frac{1}{n} \sum_{j=1}^{n} c_{ij}w_{max}(a_j),$$

where $\gamma = n/\lambda_{max}$. So we can say that the priority of the i-th alternative $w_{max}(a_i)$ is a rescaled arithmetic mean of its approximations in the form $c_{ij}w_{max}(a_j)$, where the scaling factor is γ.

Of course, the question arises as to whether such scaling does not negatively affect the ranking obtained. After all, both equations (3.4) and (3.8) are different. The answer to this question, indicating the correctness of adopting w_{max} as a ranking comes from the fact that the value of the priority of the i-th alternative makes sense only with reference to the j-th alternative. In other

---

[3] As some eigenvalues of C can be complex numbers, we can compare their modules. The Perron theorem says that λ_{max} is strictly greater than the module $|\,|$ of any other eigenvalue of C.

words, we are interested in the $w_{max}(a_i)/w_{max}(a_j)$ ratio and not in the actual value of the numerator and denominator. It is easy to see that:

$$\frac{w_{max}(a_i)}{w_{max}(a_j)} = \frac{\gamma \frac{1}{n} \sum_{r=1}^{n} c_{ir} w_{max}(a_r)}{\gamma \frac{1}{n} \sum_{r=1}^{n} c_{jr} w_{max}(a_r)},$$

and as follows,

$$\frac{w_{max}(a_i)}{w_{max}(a_j)} = \frac{\frac{1}{n} \sum_{r=1}^{n} c_{ir} w_{max}(a_r)}{\frac{1}{n} \sum_{r=1}^{n} c_{jr} w_{max}(a_r)}.$$

In other words, in the context of the ratio $w_{max}(a_i)/w_{max}(a_j)$, the scaling factor γ does not matter. Hence, for the purpose of the ranking calculation we can accept that $w_{max}(a_i)$ is, in fact, the arithmetic mean of all its approximations, which was our main postulate when we were formulating the equation (3.4).

The fact that the value $w_{max}(a_i)$ has only a relative meaning means that the particular priority must always be considered together with other ranking values. This is obviously inconvenient, because at first glance it is difficult to assess whether a given value is large or small. However, we can easily remedy this. Since scaling the priority vector w keeps the ratios $w_{max}(a_i)/w_{max}(a_j)$ unchanged, let us divide all the priority values by their sum. Hence, let the new priority function w_{ev} be defined as:

$$w_{ev}(a_i) \stackrel{df}{=} \frac{w_{max}(a_i)}{\sum_{j=1}^{n} w_{max}(a_j)} \text{ for } i = 1, \ldots, n,$$

and

$$w_{ev} = \begin{pmatrix} w_{ev}(a_1) \\ w_{ev}(a_2) \\ \vdots \\ w_{ev}(a_n) \end{pmatrix}.$$

After this operation, all the priority values sum up to 1. Therefore, an individual value $w_{ev}(a_i) \cdot 100\%$ can be interpreted as a percentage. This may help in assessing the importance of the given alternative, especially if we know the total number of alternatives.

The idea of accepting an appropriately rescaled principal eigenvector of C as the list of priority values for the alternatives $a_1, \ldots, a_n$ is called the eigenvalue method (EVM).

3.1.2 Illustrative example

Example 3. *Let us consider a simple 3×3 PC matrix*

$$C = \begin{pmatrix} 1 & 2 & 3 \\ \frac{1}{2} & 1 & 4 \\ \frac{1}{3} & \frac{1}{4} & 1 \end{pmatrix}.$$

Then we may compute that the principal eigenvector for C *is*

$$w_{max} = \begin{pmatrix} 0.806 \\ 0.558 \\ 0.193 \end{pmatrix}.$$

Thus, the EVM ranking is

$$w_{ev} = \begin{pmatrix} 0.517 \\ 0.358 \\ 0.124 \end{pmatrix}.$$

Usually, as in the above example, we calculate the principal eigenvector using software such as *Microsoft Excel*[4], *MatLab*[5], R[6], or *Mathematica*[7], then we rescale it so that all its components sum up to 1. This is due to the fact that the use of direct formulas would be rather onerous. For example, for a 3×3 PC matrix

$$C = \begin{pmatrix} 1 & c_{1,2} & c_{1,3} \\ \frac{1}{c_{2,1}} & 1 & c_{2,3} \\ \frac{1}{c_{3,1}} & \frac{1}{c_{3,2}} & 1 \end{pmatrix}$$

the direct formula for the principal eigenvalue is

$$\lambda_{max} = \delta + \frac{1}{\delta} + 1,$$

where

$$\delta = \frac{\sqrt[3]{\sqrt{\left(\frac{27c_{1,3}}{c_{1,2}c_{2,3}} + \frac{27c_{1,2}c_{2,3}}{c_{1,3}}\right)^2 - 2916} + \frac{27c_{1,2}c_{2,3}}{c_{1,3}} + \frac{27c_{1,3}}{c_{1,2}c_{2,3}}}}{3\sqrt[3]{2}}.$$

This leads us to the principal eigenvector:

$$w_{max} = \begin{pmatrix} c_{1,3}\left(\delta + \frac{1}{\delta}\right) - \dfrac{c_{1,3}\left(\frac{c_{2,3}}{c_{1,3}} + \frac{\delta + \frac{1}{\delta}}{c_{1,2}}\right)}{c_{2,3}\left(\frac{\delta + \frac{1}{\delta}}{c_{1,3}} + \frac{1}{c_{1,2}c_{2,3}}\right)} \\[4ex] \dfrac{\frac{c_{2,3}}{c_{1,3}} + \frac{\delta + \frac{1}{\delta}}{c_{1,2}}}{\frac{\delta + \frac{1}{\delta}}{c_{1,3}} + \frac{1}{c_{1,2}c_{2,3}}} \\[4ex] 1 \end{pmatrix}.$$

and, finally, we may write down the appropriately rescaled priority vector w_{ev}.

---

[4]Proprietary software provided by *Microsoft Corporation*.
[5]Proprietary software provided by *MathWorks Inc.*
[6]R language – free and open software language created mainly for statistical calculations.
[7]Proprietary software provided by *Wolfram Research*.

3.2 Geometric mean method

3.2.1 Idea of the method

In the previous section, the EVM was presented. It is based on the observation that $w(a_i)$ can be approximated by the weighted mean of components in the form $c_{ij}w(a_j)$. It is possible to ask, however, why the geometric mean is not used instead of the arithmetic mean. The first to propose the use of geometric mean were Crawford and Williams [44]. It is interesting that this method, called GMM – Geometric Mean Method, proved to be both easier to calculate and potentially better in terms of the compliance of results with the original matrix.

Thus, as before, starting from the assumption that $w(a_i) \approx c_{ij}w(a_j)$, we may request that $w(a_i)$ be given as a geometric mean of all the elements in the form $c_{ij}w(a_j)$, where $j = 1, \ldots, n$ i.e.,

$$w(a_i) = \sqrt[n]{\prod_{j=1}^{n} c_{ij}w(a_j)} \quad \text{for} \quad i = 1, \ldots, n. \tag{3.9}$$

The equation (3.9) looks more complicated than their additive counterpart (3.4). Fortunately, due to the fact that we are interested in the ratio $w(a_i)/w(a_j)$, the formula (3.9) is significantly simplified. Let us write (3.9) as:

$$w(a_i) = \sqrt[n]{\prod_{k=1}^{n} c_{ik}} \sqrt[n]{\prod_{j=k}^{n} w(a_k)}.$$

The ratio $w(a_i)/w(a_j)$ can be written as:

$$\frac{w(a_i)}{w(a_j)} = \frac{\sqrt[n]{\prod_{k=1}^{n} c_{ik}} \sqrt[n]{\prod_{j=k}^{n} w(a_k)}}{\sqrt[n]{\prod_{k=1}^{n} c_{jk}} \sqrt[n]{\prod_{k=1}^{n} w(a_k)}},$$

i.e.,

$$\frac{w(a_i)}{w(a_j)} = \frac{\sqrt[n]{\prod_{k=1}^{n} c_{ik}}}{\sqrt[n]{\prod_{k=1}^{n} c_{jk}}}. \tag{3.10}$$

This clearly indicates that, for the purpose of calculating the relative priority of the i-th alternative, it is enough to adopt the geometric mean of the i-th row of C. Thus, instead of (3.9), we may accept as the unscaled geometric mean priority of the i-th alternative the value:

$$w_{gmu}(a_i) \stackrel{df}{=} \sqrt[n]{\prod_{k=1}^{n} c_{ik}}. \tag{3.11}$$

The formula is easy to calculate and does not require the use of specialized software as the EVM does. Of course, similarly to the case of EVM, it is better to scale the priority vector to 1. Therefore, ultimately (3.11) takes the form:

$$w_{gm}(a_i) \stackrel{df}{=} \frac{w_{gmu}(a_i)}{\sum_{j=1}^{n} w_{gmu}(a_j)}, \qquad (3.12)$$

i.e.,

$$w_{gm}(a_i) \stackrel{df}{=} \frac{\sqrt[n]{\prod_{k=1}^{n} c_{ik}}}{\sum_{j=1}^{n} \sqrt[n]{\prod_{k=1}^{n} c_{jk}}}, \qquad (3.13)$$

and

$$w_{gm} \stackrel{df}{=} \begin{pmatrix} w_{gm}(a_1) \\ w_{gm}(a_2) \\ \vdots \\ w_{gm}(a_n) \end{pmatrix}. \qquad (3.14)$$

The method of using formulas (3.13) and (3.14) to calculate the ranking based on a PC matrix is called GMM.

3.2.2   Illustrative example

Example 4. *Let us consider a 5×5 PC matrix corresponding to the pairwise judgments of five alternatives $a_1, \ldots, a_5$:*

$$C = \begin{pmatrix} 1 & 4 & \frac{5}{3} & \frac{6}{7} & 2 \\ \frac{1}{4} & 1 & \frac{6}{5} & \frac{5}{9} & \frac{3}{5} \\ \frac{3}{5} & \frac{5}{6} & 1 & \frac{1}{3} & \frac{1}{2} \\ \frac{7}{6} & \frac{9}{5} & 3 & 1 & \frac{4}{9} \\ \frac{1}{2} & \frac{5}{3} & 2 & \frac{9}{4} & 1 \end{pmatrix}.$$

Thus, the geometric means of rows are:

$$
\begin{aligned}
\sqrt[5]{1 \cdot 4 \cdot \tfrac{5}{3} \cdot \tfrac{6}{7} \cdot 2} &= \sqrt[5]{\tfrac{80}{7}} \\
\sqrt[5]{\tfrac{1}{4} \cdot 1 \cdot \tfrac{6}{5} \cdot \tfrac{5}{9} \cdot \tfrac{3}{5}} &= \sqrt[5]{\tfrac{1}{10}} \\
\sqrt[5]{\tfrac{3}{5} \cdot \tfrac{5}{6} \cdot 1 \cdot \tfrac{1}{3} \cdot \tfrac{1}{2}} &= \sqrt[5]{\tfrac{1}{12}} \, , \\
\sqrt[5]{\tfrac{7}{6} \cdot \tfrac{9}{5} \cdot 3 \cdot 1 \cdot \tfrac{4}{9}} &= \sqrt[5]{\tfrac{14}{5}} \\
\sqrt[5]{\tfrac{1}{2} \cdot \tfrac{5}{3} \cdot 2 \cdot \tfrac{9}{4} \cdot 1} &= \sqrt[5]{\tfrac{15}{4}}
\end{aligned}
$$

and finally, the priority vector is:

$$
\begin{pmatrix} w_{gm}(a_1) \\ w_{gm}(a_2) \\ w_{gm}(a_3) \\ w_{gm}(a_4) \\ w_{gm}(a_5) \end{pmatrix} = \begin{pmatrix} \dfrac{\sqrt[5]{\frac{80}{7}}}{\sqrt[5]{\frac{80}{7}}+\sqrt[5]{\frac{1}{10}}+\sqrt[5]{\frac{1}{12}}+\sqrt[5]{\frac{14}{5}}+\sqrt[5]{\frac{15}{4}}} \\ \dfrac{\sqrt[5]{\frac{1}{10}}}{\sqrt[5]{\frac{80}{7}}+\sqrt[5]{\frac{1}{10}}+\sqrt[5]{\frac{1}{12}}+\sqrt[5]{\frac{14}{5}}+\sqrt[5]{\frac{15}{4}}} \\ \dfrac{\sqrt[5]{\frac{1}{12}}}{\sqrt[5]{\frac{80}{7}}+\sqrt[5]{\frac{1}{10}}+\sqrt[5]{\frac{1}{12}}+\sqrt[5]{\frac{14}{5}}+\sqrt[5]{\frac{15}{4}}} \\ \dfrac{\sqrt[5]{\frac{14}{5}}}{\sqrt[5]{\frac{80}{7}}+\sqrt[5]{\frac{1}{10}}+\sqrt[5]{\frac{1}{12}}+\sqrt[5]{\frac{14}{5}}+\sqrt[5]{\frac{15}{4}}} \\ \dfrac{\sqrt[5]{\frac{15}{4}}}{\sqrt[5]{\frac{80}{7}}+\sqrt[5]{\frac{1}{10}}+\sqrt[5]{\frac{1}{12}}+\sqrt[5]{\frac{14}{5}}+\sqrt[5]{\frac{15}{4}}} \end{pmatrix} = \begin{pmatrix} 0.3015 \\ 0.1169 \\ 0.1127 \\ 0.2276 \\ 0.2413 \end{pmatrix}.
$$

Thus, according to the created ranking, the most important (the most pre-ferred) alternative is a_1 with the priority $w_{gm}(a_1) = 0.3015$. The second, third, fourth and fifth places are taken by a_5, a_4, a_2 and a_3 correspondingly.

3.2.3 Is GMM optimal?

Where the same problem can be solved in many different ways, a natural question arises as to which way is the best. The same question can be asked for EVM and GMM. In their work, *Crawford and Williams* [44] presented an important argument for using the GMM method. This argument is the fact that GMM minimizes the total error of the method, which may indicate its optimality. Let's take a closer look at this problem.

Since it is natural to expect that c_{ij} equals $w(a_i)/w(a_j)$ then also:

$$
\frac{c_{ij}w(a_j)}{w(a_i)} = 1, \quad \text{for } i,j = 1,\ldots,n.
$$

Hence, we can write the above condition in the log form:

$$
\ln c_{ij} - \ln \frac{w(a_i)}{w(a_j)} = 0.
$$

Since, due to the inconsistency, $w(a_i)/w(a_j)$ only approximates c_{ij} then also

$$
\ln c_{ij} - \ln \frac{w(a_i)}{w(a_j)} \neq 0.
$$

Let us define the square of the difference between $\log c_{ij}$ and $\log w(a_i)/w(a_j)$ as the error e_{ij}:

$$
e_{ij} \stackrel{df}{=} \left(\ln c_{ij} - \ln \frac{w(a_i)}{w(a_j)} \right)^2.
$$

Of course the smaller e_{ij} the better[8]. Therefore, in the context of the priority-deriving method w and the PC matrix C the errors e_{ij}, as defined above, can

---

[8]In fact, we are interested in making $\log \frac{w(a_i)}{w(a_j)}$ as close as possible to $\log c_{ij}$. However, sometimes $\log c_{ij}$ can be larger, other times $\log \frac{w(a_i)}{w(a_j)}$, therefore it is convenient to consider the squared difference denoted as e_{ij}. The smaller e_{ij} the closer $\log \frac{w(a_i)}{w(a_j)}$ and $\log c_{ij}$.

be used for construction of an optimality criterion. To this end, let us define
the total error $\mathcal{E}$ as the sum of errors e_{ij}

$$\mathcal{E} \overset{df}{=} \sum_{i=1}^{n} \sum_{j=1}^{n} e_{ij}. \tag{3.15}$$

Since the formula (3.15) takes into account all the elements e_{ij} we may claim
that the optimal priority-deriving method should minimize the total error
$\mathcal{E}$. Surprisingly, GMM has this property. To prove it, let us define $\mathcal{E}$ as the
function of w,

$$\mathcal{E}(w(a_1), \ldots, w(a_n)) \overset{df}{=} \sum_{i=1}^{n} \sum_{j=1}^{n} \left(\ln c_{ij} - \ln \frac{w(a_i)}{w(a_j)} \right)^2.$$

By replacing $u_i \overset{df}{=} \ln w(a_i)$ we transform $\mathcal{E}$ to $\widehat{\mathcal{E}}$:

$$\widehat{\mathcal{E}}(u_1, \ldots, u_n) \overset{df}{=} \sum_{i=1}^{n} \sum_{j=1}^{n} (\ln c_{ij} - u_i + u_j)^2.$$

Thus, the proposition of GMM optimality takes the form of the following
theorem:

Theorem 1. *GMM priority vector minimizes* $\widehat{\mathcal{E}}$.

Proof. As we know, the function $\mathcal{E}$ reaches a critical point when its first partial
derivatives get 0. This point is maximal if the second derivative matrix of
$\mathcal{E}$ is positive semidefinite [26, 146]. Therefore, the following proof comprises
the calculation of the first and second derivative of the function $\widehat{\mathcal{E}}$ where
$u_1, \ldots, u_n$ are variables of $\widehat{\mathcal{E}}$. On the basis of these calculations, we will be
able to determine what the optimal y_i should be.

Let us compare a vector of partial derivatives $\mathcal{E}$ to zero:

$$\begin{pmatrix} \frac{\partial \widehat{\mathcal{E}}}{\partial u_1} \\ \frac{\partial \widehat{\mathcal{E}}}{\partial u_2} \\ \vdots \\ \frac{\partial \widehat{\mathcal{E}}}{\partial u_n} \end{pmatrix} = 0 \tag{3.16}$$

where

$$\frac{\partial \widehat{\mathcal{E}}}{\partial u_i} = \frac{\partial}{\partial u_i} \sum_{i=1}^{n} \sum_{j=1}^{n} (\ln c_{ij} - u_i + u_j)^2. \tag{3.17}$$

Thus, we need to solve

$$\frac{\partial}{\partial u_i} \sum_{k=1}^{n} \sum_{j=1}^{n} (\ln c_{ij} - u_k + u_j)^2 = 0, \tag{3.18}$$

for arbitrarily selected $i = 1, \ldots, n$. By calculating the derivative (3.17), the formula (3.18) is transformed to:

$$\sum_{j=1,i\neq j}^{n} 2\left(\ln \frac{c_{j,i}}{c_{i,j}}\right) + 4(n-1)u_i - \sum_{j=1,j\neq i}^{n} 4u_j = 0.$$

Then, due to the reciprocity

$$\sum_{j=1,i\neq j}^{n} 2\left(\ln \frac{1}{c_{ij}^2}\right) + 4(n-1)u_i - \sum_{j=1,j\neq i}^{n} 4u_j = 0,$$

i.e.,

$$\sum_{j=1,i\neq j}^{n} 4\left(\ln \frac{1}{c_{ij}}\right) + 4(n-1)u_i - \sum_{j=1,j\neq i}^{n} 4u_j = 0,$$

and as follows:

$$\frac{\partial \widehat{\mathcal{E}}}{\partial u_i} = 4\left(\sum_{j=1,i\neq j}^{n} \ln \frac{1}{c_{ij}} + (n-1)u_i - \sum_{j=1,j\neq i}^{n} u_j\right) = 0. \tag{3.19}$$

By adding u_i to both sides of (3.19) we obtain:

$$nu_i = \sum_{j=1}^{n} u_j - \sum_{j=1}^{n} \ln \frac{1}{c_{ij}}.$$

The above equation can be written in the form

$$\ln w(a_i) = \frac{1}{n}\left(\sum_{j=1}^{n} \ln w(a_j) + \sum_{j=1}^{n} \ln c_{ij}\right).$$

Since $w(a_i) \neq 0$ for $i = 1, \ldots, n$ then the above equation holds if and only if

$$\ln w(a_i) = \frac{1}{n} \sum_{j=1}^{n} \ln c_{ij} w(a_j).$$

Due to the properties of the logarithm function

$$\ln w(a_i) = \frac{1}{n} \ln \prod_{j=1}^{n} c_{ij} w(a_j),$$

and

$$\ln w(a_i) = \ln \sqrt[n]{\prod_{j=1}^{n} c_{ij} w(a_j)}.$$

Since the logarithm function is injective then

$$w(a_i) = \sqrt[n]{\prod_{j=1}^{n} c_{ij} w(a_j)}. \tag{3.20}$$

This means that $\widehat{\mathcal{E}}$ reaches its critical point wherever its arguments meet the condition (3.20). However, as expected, (3.20) is an equation defining the method of calculating the priority of the i-th alternative (3.11). This means that indeed w_{gm} is expected to be the minimum of $\mathcal{E}$. To find out about this definitively, let us consider the second derivative of the error function. Hence,

$$\widehat{\mathcal{E}}(y_1, \ldots, y_n) \overset{df}{=} \sum_{i=1}^{n} \sum_{j=1}^{n} \left(\ln c_{ij} - y_i + y_j \right)^2.$$

We have, then

$$H_{\widehat{\mathcal{E}}} = \begin{pmatrix} \frac{\partial^2 \widehat{\mathcal{E}}}{\partial^2 x_1} & \frac{\partial^2 \widehat{\mathcal{E}}}{\partial x_1 \partial x_2} & \cdots & \frac{\partial^2 \widehat{\mathcal{E}}}{\partial x_1 \partial x_n} \\ \frac{\partial^2 \widehat{\mathcal{E}}}{\partial x_2 \partial x_1} & \frac{\partial^2 \widehat{\mathcal{E}}}{\partial^2 x_2} & \cdots & \vdots \\ \vdots & \cdots & \ddots & \vdots \\ \frac{\partial^2 \widehat{\mathcal{E}}}{\partial x_n \partial x_1} & \cdots & \cdots & \frac{\partial^2 \widehat{\mathcal{E}}}{\partial^2 x_n} \end{pmatrix},$$

which boils down to

$$H_{\widehat{\mathcal{E}}} = \begin{pmatrix} 4(n-1) & -4 & \cdots & -4 \\ -4 & 4(n-1) & \cdots & \vdots \\ \vdots & \cdots & \ddots & \vdots \\ -4 & \cdots & \cdots & 4(n-1) \end{pmatrix},$$

$\widehat{\mathcal{E}}$ is convex if $H_{\widehat{\mathcal{E}}}$ is positive semidefinite [146, 20, p. 71], i.e., if for any real vector $x \in \mathbb{R}^n$ it holds that $x^T H_{\mathcal{E}} x \geq 0$. Thus, let us calculate the scalar $x^T H_{\mathcal{E}} x$.

$$x^T H_{\mathcal{E}} x = 4(n-1) \sum_{i=1}^{n} x_i^2 - 8 \sum_{i=1}^{n-1} \sum_{j=i+1}^{n} x_i x_j.$$

Then,

$$x^T H_{\mathcal{E}} x = \frac{1}{4} \left((n-1) \sum_{i=1}^{n} x_i^2 - 2 \sum_{i=1}^{n-1} \sum_{j=i+1}^{n} x_i x_j \right).$$

The expression on the right side of the above formula can be written in the more concise form:

$$x^T H_{\mathcal{E}} x = \frac{1}{4} \sum_{i=1}^{n-1} \sum_{j=i+1}^{n} (x_i - x_j)^2. \tag{3.21}$$

Since $(x_i - x_j)^2 \geq 0$ for any $x_i, x_j \in \mathbb{R}$ then also the sum on the right side of (3.21) is non negative. Therefore, finally, $x^T H_{\widehat{\mathcal{E}}} x \geq 0$, hence $H_{\widehat{\mathcal{E}}}$ is positive semidefinite for any values $w(a_1), \ldots, w(a_n)$. Thus, $\widehat{\mathcal{E}}$ is convex on $\mathbb{R}^n$ [20, p. 71], and (3.14) is the minimum of $\widehat{\mathcal{E}}$.

$\square$

3.3 Optimization methods

In addition to the most popular EVM and GMM, there is also an interesting group of methods based on numerical optimization. [72, 40, 21]. *Golany* and *Kress* call them the extremal methods [72]. *Choo* and *Wedley* refer to them as the distance minimization estimation methods. All of them are based on the observation that c_{ij} should be as close to $w(a_i)/w(a_j)$ as possible. This fact serves them to formulate an optimization problem, which is then solved using numerical analysis. In the case of GMM, it was difficult at first glance to decide whether and with respect to which criterion the method is optimal. In the case of the considered methods, they themselves are determined by the optimality criterion. Thus, the obtained results are (in some sense) optimal.

3.3.1 Least squares method

An example of this class of methods is the Least Squares Method (LSM). This method consists of solving the following minimization problem:

$$\min \quad \mathcal{E} = \sum_{i=1}^{n} \sum_{j=1}^{n} \left(c_{ij} - \frac{w(a_i)}{w(a_j)} \right)^2,$$

subject to

$$\sum_{i=1}^{n} w(a_i) = 1,$$

and

$$w(a_i) > 0, \quad \text{for } i = 1, \ldots, n.$$

The LSM problem is difficult to solve as the objective function is nonlinear, usually non-convex and no unique solution exists [21]. The Newton–Kantorovich (N–K) method of approximation can be used to solve LSM, but a proper initial point is desired [59].

3.3.2   Weighted least squares method

Similar to LSM is the Weighted Least Squares Method (WLSM) [72, 40]. Since the difference $c_{ij} - w(a_i)/w(a_j)$ should tend to 0, then the expression

$$w(a_j)\,(c_{ij} - w(a_i)/w(a_j)) \approx 0$$

and as follows,

$$c_{ij}w(a_j) - w(a_i) \approx 0.$$

Hence, finding the priority vector boils down to solving the following optimization problem:

$$\min \quad \mathcal{E} = \sum_{i=1}^{n} \sum_{j=1}^{n} (c_{ij}w(a_j) - w(a_i))^2, \tag{3.22}$$

subject to

$$\sum_{i=1}^{n} w(a_i) = 1, \tag{3.23}$$

and

$$w(a_i) > 0, \quad \text{for } i = 1, \ldots, n.$$

The advantage of the above problem is the existence of a relatively simple solution. By calculating partial derivatives of (3.22) and equating it to 0 we get the linear equation system:

$$\frac{\partial \mathcal{E}}{\partial w(a_i)} = 2 \sum_{j=1, j\neq i}^{n} c_{ji}\,(w(a_i)c_{ji} - w(a_j)) - (w(a_j)c_{ij} - w(a_i)) = 0$$

for $i = 1, \ldots, n$. This leads [40] to the matrix equation

$$Bw = 0, \tag{3.24}$$

where $B = [b_{ij}]$

$$b_{ij} = \begin{cases} -(c_{ij} + c_{ji}) & \text{if } i \neq j \\ (n-2) + \sum_{k=1, k\neq i}^{n} c_{ki}^2 & \text{if } i = j \end{cases}$$

and (3.23) holds. Solving (3.24) provides the priority vector candidate.

To make our reasoning complete, let us check the Hessian matrix $H_{\mathcal{E}} = [h_{ij}]$. Using elementary transformations we obtain:

$$h_{ij} = \begin{cases} -2\,(c_{ij} + c_{ji}) & \text{if } i \neq j \\ 2(n-1) + 2\sum_{k=1, k\neq i}^{n} c_{ki} & \text{if } i = j \end{cases}.$$

In order to check whether $H_{\mathcal{E}}$ is positive semidefinite, let us compute $x^T H_{\mathcal{E}} x$ for any real vector $x \in \mathbb{R}^n$. Thus, it is easy to verify that

$$x^T H_{\mathcal{E}} x = 2(n-1) \sum_{j=1}^{n} x_j^2 + 2 \sum_{i=1}^{n} \sum_{j=1,j\neq i}^{n} (x_i c_{ij})^2 - 4 \sum_{i=1}^{n} \sum_{j=1,j\neq i}^{n} x_i x_j (c_{ij} + c_{ji}).$$

This, however, is equivalent to:

$$x^T H_{\mathcal{E}} x = 2 \sum_{i=1}^{n-1} \sum_{j=i+1}^{n} (x_i - x_j c_{ij})^2 + (x_j - x_i c_{ji})^2.$$

Because on the right side of the above equation there is the sum of squares, it is non-negative, hence, $x^T H_{\mathcal{E}} x \geq 0$. Due to the convexity of $\mathcal{E}$, the solution of (3.24) is the desired minimum.

3.3.3   Logarithmic least squares method

The Logarithmic Least Squares Method (LLSM) is based on the observation $c_{ij} \approx w(a_i)/w(a_j)$. It implies that $(\ln c_{ij} - \ln w(a_i) + \ln w(a_j))^2 \approx 0$. Hence, the optimization criterion takes the form:

$$\min \quad \mathcal{E} = \sum_{i=1}^{n} \sum_{j=1}^{n} (\ln c_{ij} - \ln w(a_i) + \ln w(a_j))^2.$$

subject to

$$\sum_{i=1}^{n} w(a_i) = 1,$$

and

$$w(a_i) > 0 \text{ for } i = 1, \ldots, n.$$

Of course, as was shown in (Section 3.2.3), LLSM is equivalent to GMM. For this reason, usually, this method is not considered independently of GMM.

3.3.4   Least worst squares method

The observation that $c_{ij} \approx w(a_i)/w(a_j)$ leads to the conclusion that $\max_{i\neq j} (c_{ij} - w(a_i)/w(a_j))^2 \approx 0$. This idea underlies [40] the Least Worst Squares Method (LWSM). The LWSM consists of finding a solution to the following optimization problem:

$$\min \quad \mathcal{E} = \max_{i\neq j} (c_{ij} - w(a_i)/w(a_j))^2,$$

subject to

$$\sum_{i=1}^{n} w(a_i) = 1,$$

and
$$w(a_i) > 0 \text{ for } i = 1, \ldots, n.$$

Similarly to LSM, the above nonlinear problem has no direct solution and it is difficult to solve [40].

3.3.5   Weighted least worst squares method

Similarly to the case of WLSM, we may observe that for the given PC matrix C and priority vector w it should hold that $c_{ij}w(a_j) - w(a_i) \approx 0$. As this condition is true, it holds for any $i, j = 1, \ldots, n$ then in particular also $max_{i \neq j} (c_{ij}w(a_j) - w(a_i))^2 \approx 0$. This leads to the Weighted Least Worst Squares Method (WLWSM) given in the form of the following optimization criterion:

$$\min \quad \mathcal{E} = max_{i \neq j} \left(c_{ij}w(a_j) - w(a_i) \right)^2,$$

subject to

$$\sum_{i=1}^{n} w(a_i) = 1,$$

and

$$w(a_i) > 0 \text{ for } i = 1, \ldots, n.$$

In contrast to WLSM, the WLWSM method does not have any direct solution [40] and has to be solved using iterative approximation methods.

3.3.6   Weighted least absolute error method

The Weighted Least Absolute Error Method (WLAEM) is similar to the WLSM approach [40]. The difference is the way in which the distance between $c_{ij}w(a_j)$ and $w(a_i)$ is calculated. In WLSM, $c_{ij}w(a_j) - w(a_i)$ is squared, while in WLAEM we are interested in the absolute value of this difference. Hence, the optimization criterion is given as:

$$\min \quad \mathcal{E} = \sum_{i=1}^{n} \sum_{j=1}^{n} |c_{ij}w(a_j) - w(a_i)|,$$

subject to

$$\sum_{i=1}^{n} w(a_i) = 1,$$

and

$$w(a_i) > 0 \text{ for } i = 1, \ldots, n.$$

Fortunately, the above problem can be transformed into an equivalent linear programing problem:

$$\min \quad \sum_{i,j=1, i\neq j}^{n} P_{ij} + \sum_{i,j=1, i\neq j}^{n} M_{ij},$$

subject to

$$c_{ij}w(a_j) - w(a_i) - P_{ij} + M_{ij} = 0 \quad \text{for } 1 \leq i \neq j \leq n,$$
$$w(a_i) \geq 1 \quad \text{for } i = 1, \ldots, n,$$
$$P_{ij} \geq 0, \quad M_{ij} \geq 0, \quad \text{for } i, j = 1, \ldots, n \text{ and } i \neq j,$$
$$w(a_1) + \ldots + w(a_n) = 1,$$

where P_{ij} and M_{ij} are called deviation variables. Solving the above linear program provides the desired solution.

3.3.7 Weighted least worst absolute error method

In the same way as WLAEM is similar to WLSM, the Weighted Least Worst Absolute Error Method (WLWAEM) is similar to WLWSM [40]. Thus, the optimization criterion is given as:

$$\min \quad \mathcal{E} = \max_{i \neq j} |c_{ij}w(a_j) - w(a_i)|,$$

providing that

$$\sum_{i=1}^{n} w(a_i) = 1.$$

This problem also has its linear programing representation, which looks as follows:

$$\min \quad Z,$$

subject to

$$Z \geq c_{ij}w(a_j) - w(a_i) \quad \text{for } 1 \leq i \neq j \leq n,$$
$$Z \geq -c_{ij}w(a_j) + w(a_i) \quad \text{for } 1 \leq i \neq j \leq n,$$
$$w(a_i) \geq 1 \quad \text{for } i = 1, \ldots, n,$$
$$w(a_1) + \ldots + w(a_n) = 1.$$

Solving the above linear program leads to the desired priority vector.

3.4　Weighted column sum methods

3.4.1　Simple column sum method

One of the most simple priority-deriving methods is to compute the average of rows of the PC matrix [201]. The method is also called the Simple Column Sum Method (SCSM) [40]. Hence, the priority vector is given as:

$$
\begin{pmatrix} w(a_1) \\ \vdots \\ \vdots \\ w(a_n) \end{pmatrix} = \begin{pmatrix} \frac{1}{n} \sum_{j=1}^n c_{1j} \\ \vdots \\ \vdots \\ \frac{1}{n} \sum_{j=1}^n c_{nj} \end{pmatrix}.
$$

In this approach, every column is scaled by the same $1/n$ weight, i.e., every priority $w(a_i)$ is the arithmetic mean of the relative priority of a_i with respect to all other alternatives.

3.4.2　Simple scaled column sum method

One of the drawbacks of the above method is the fact that different columns may not have commensurate units [40]. The answer to this problem is their scaling, so that every column sums up to 1 [40]. This gives rise to the Simple Scaled Column Sum Method (SSCSM) in which the priority vector is given as:

$$
\begin{pmatrix} w(a_1) \\ \vdots \\ \vdots \\ w(a_n) \end{pmatrix} = \begin{pmatrix} \sum_{j=1}^n \frac{c_{1j}}{\sum_{k=1}^n c_{kj}} \\ \vdots \\ \vdots \\ \sum_{j=1}^n \frac{c_{nj}}{\sum_{k=1}^n c_{kj}} \end{pmatrix}. \tag{3.25}
$$

It is interesting that if C is consistent (see Sec. 1.2.6), and as follows if $w(a_i) = c_{ij}w(a_j)$ (see Theorem 4) and both: the priority vector w and every column of C are scaled so that they sum up to one, then

$$
w(a_i) = c_{ij}, \tag{3.26}
$$

for every $i, j = 1, \ldots, n$. To verify that, let us consider unscaled w and C with unscaled columns. According to (3.26), we may expect that

$$
\frac{w(a_i)}{\sum_{k=1}^n w(a_k)} = \frac{c_{ij}}{\sum_{k=1}^n c_{kj}}, \tag{3.27}
$$

for every $i, j = 1, \ldots, n$. The above holds if and only if

$$\frac{c_{ij}w(a_j)}{\sum_{k=1}^{n}w(a_k)} = \frac{c_{ij}}{\sum_{k=1}^{n}\frac{w(a_k)}{w(a_j)}},$$

hence

$$c_{ij}w(a_j)\sum_{k=1}^{n}\frac{w(a_k)}{w(a_j)} = c_{ij}\sum_{k=1}^{n}w(a_k),$$

which leads to

$$c_{ij}\sum_{k=1}^{n}w(a_k) = c_{ij}\sum_{k=1}^{n}w(a_k),$$

which is undoubtedly true. The above reasoning shows that, in the case of a consistent PC matrix, the SSCSM produces the priority vector n times greater than its scaled counterpart in EVM or GMM. However, because we are ultimately interested in ratios $w(a_i)/w(a_j)$, the scaling factor does not matter.

3.4.3 A cosine maximization method

The cosine maximization method (CMM) [107] is the idea of connecting optimization models and properties of the columns of the PC matrix C. As such, it could easily be placed together with optimization methods as well as in a separate section. However, if we take a look closer at it, we can see the same principles that have formed CMM, SCSM and SSCSM. The last two of these methods are based on the observation that the priority vector w should be similar to every column of the matrix $C = [C_1, \ldots, C_n]$ with respect to some scaling factor α. I.e., $w \approx \alpha C_i$ for $i = 1, \ldots, n$, where C_i means the i-th column of C. In particular, as shown above, if w and C_i are scaled so that they sum up to one, and C is consistent, then $w = C_i$ for $i = 1, \ldots, n$. The fact that $w \approx \alpha C_i$ translates to the observation that the cosine between two vectors $w, C_i \in \mathbb{R}^n$ is close to one, i.e., $\cos(w, C_i) \approx 1$ for every $i = 1, \ldots, n$, where

$$\cos(w, C_i) = \frac{\sum_{k=1}^{n}w(a_k)c_{ik}}{\sqrt{\sum_{k=1}^{n}w^2(a_k)}\sqrt{\sum_{k=1}^{n}c_{ki}^2}}. \tag{3.28}$$

In particular, if w and C_i differ from each other only by a constant factor α, i.e., it holds that $w = \alpha C_i$, then $\cos(w, C_i) = 1$. Since the closer to 1 that $\cos(w, C_i)$ is the better, then CMM proposes to find a w that will be as close as possible to each C_i. This postulate leads to the formulation of the following optimization model:

$$\max \sum_{i=1}^{n}\cos(w, C_i), \tag{3.29}$$

subject to

$$\sum_{i=1}^{n} w(a_i) = 1,$$

and

$$w(a_i) > 0, \text{ and } i = 1, \dots, n.$$

By adopting

$$\widehat{w}(a_i) = \frac{w(a_i)}{\sqrt{\sum_{k=1}^{n} w^2(a_k)}}, \text{ for } i = 1, \dots, n,$$

and

$$b_{ij} = \frac{c_{ij}}{\sum_{k=1}^{n} c_{kj}^2}, \text{ for } i, j = 1, \dots, n,$$

we may reformulate the above problem to the equivalent one:

$$\max \sum_{i=1}^{n} \cos(w, C_i) = \sum_{j=1}^{n} \sum_{i=1}^{n} b_{ij} \widehat{w}(a_i) = \sum_{i=1}^{n} \left(\sum_{j=1}^{n} b_{ij} \right) \widehat{w}(a_i), \quad (3.30)$$

subject to

$$\sum_{i=1}^{n} \widehat{w}^2(a_i) = 1,$$

and

$$w(a_i) \geq 0, \text{ for } i = 1, \dots, n.$$

Fortunately, the latter optimization model has an easy to calculate solution [107, Theorem 2]. It is the vector $w^* = [w^*(a_1), \dots, w^*(a_n)]^T$ where

$$w^*(a_i) = \frac{\sum_{j=1}^{n} b_{ij}}{\sqrt{\sum_{k=1}^{n} \left(\sum_{j=1}^{n} b_{kj} \right)^2}}.$$

In particular [107, Theorem 3] when C is consistent then:

$$w^*(a_j) = \frac{1}{\sum_{k=1}^{n} c_{kj}}. \quad (3.31)$$

Note that (3.31) implies $\sum_{k=1}^{n} w^*(a_k) = 1$, thus w^* is properly scaled right away.

3.4.4   Other methods

Of course, the methods described above do not exhaust the list of methods for calculating the ranking of alternatives. For example, approximating a pairwise comparison matrix by another consistent matrix can also be considered as a prioritization method (note that as the vector of priorities of a consistent matrix we may adopt any of its columns) [66, 102]. Different priority-deriving methods for AHP can be found in various review works [78, 88, 40].

3.5   Toward comparing prioritization methods

All the presented priority-deriving methods are based on certain heuristics. For example, EVM assumes that $w(a_i) = \sum_{j=1}^{n} c_{ij} w(a_j)$. Other methods formulate an optimization problem. Their solution provides the desired priority vector. Hence, the rankings obtained using optimization methods are optimal with respect to the given criteria. This situation may lead to some confusion and doubts about which method to choose. Of course, it would not be a bad idea to adopt a method allowing optimal selection of the best decision-making method. This method should also be optimal, but how should it be selected? Our considerations lead to looping, similarly to the popular *Plutarch's* causality dilemma: "which came first: the bird or the egg?"[144]. Ultimately, if we are making decisions, we have to decide on some method of decision making. The problem with the choice of the optimal method of decision making was noticed for the first time by *Triantaphyllou* and *Mann* [192]. They observed that different decision-making methods lead to different results, even though they use the same data. The answer concerning which method to choose, i.e., which is the best one, may lead to an appropriate decision-making method. However, it also needs to be optimal. Here the loop closes. They called their observation the decision-making paradox [192].

For this reason, it is quite difficult to indicate the one and only best method of decision making. This does not mean, however, that one can not compare methods of making decisions with one another. In the literature, we can find the criteria that can be used to assess the quality of the priority vectors obtained. These criteria are not suitable for the construction of a ranking method. Therefore, there is no situation that one chosen method will be "by definition" the best in the context of a given criterion. Of course, the meaning of these criteria is subjective. One person may attach more importance to a given criterion, others less. Therefore, definitive judgments and opinions can not be made on this basis. However, it is worth taking the criteria into account.

3.5.1   Minimum violations criterion

Very often, quantitative pairwise comparison is used qualitatively. This means that, although the priority vector is made up of real numbers for the decision maker, only the alternative with the highest priority matters. In this approach, the actual weights of alternatives do not matter but only their ordering from the smallest to the largest. The criterion of minimum violations seems to meet the needs of people who treat the PC method qualitatively. A violation, or element preference reversal, occurs when the i-th alternative is preferred to the j-th alternative in direct comparisons, i.e., when $c_{ij} > 1$, but the calculated priority of the i-th alternative is smaller than the priority of the j-th alternative, i.e., when $w(a_i) < w(a_j)$ [72, 180]. Some violations, however, seem to be less serious than others. For example, $c_{ij} = 1$ while $w(a_i) > w(a_j)$ may be considered as a smaller violation of correspondence between the direct assessment and the result of the ranking than if $c_{ij} < 1$ and $w(a_i) > w(a_j)$. Hence, violations can be divided into less and more important ones.

Of course, it is natural to expect that the optimal priority-deriving method minimizes the number of violations. These observations have resulted in the definition of the *MVC* criterion based on the number of violations. That is:

$$MVC(m) = \sum_{i=1}^{n} \sum_{j=1}^{n} I_{ij}(m),$$

where m means the priority-deriving method, and

$$I_{ij}(m) = \begin{cases} 1 & \text{if } w_m(a_i) > w_m(a_j) \text{ and } c_{ij} < 1, \\ \frac{1}{2} & \text{if } w_m(a_i) \neq w_m(a_j) \text{ and } c_{ij} = 1, \\ & \text{or } w_m(a_i) = w_m(a_j) \text{ and } c_{ij} \neq 1, \\ 0 & \text{otherwise.} \end{cases}$$

3.5.2   Conformity criterion

Very often, when listening to opinions on a given topic, we may come to the conclusion that "the truth lies somewhere in the middle." Both have arguments in their favor and it is difficult to indicate a decisive winner in the debate. What should be done then? Probably the best solution is to seek a compromise that reconciles the arguments of all the parties. This is at the heart of the conformity criterion [72]. According to it, the prioritization method is better when the calculated priority values are closer to the average of all the considered methods. The measure is computed as follows:

$$CC(m) = \sum_{i=1}^{n} |w_m(a_i) - \bar{w}(a_i)|, \quad \text{for } m = 1, \dots, T,$$

where T is the number of methods considered, and

$$\bar{w}(a_i) = \frac{1}{T} \sum_{m=1}^{T} w_m(a_i).$$

3.5.3   Other criteria

There are also other criteria, such as total deviation, robustness, invariance to transposition [72], efficiency [22] and others. One may test them using the Monte Carlo approach and, on this basis, evaluate various methods. The results of such experiments can be found in [72, 40]. These results can help to choose the calculation method, but the subjective decision of choice will ultimately always belong to man.

Chapter 4

Prioritization Methods for Incomplete PC Matrices

To create a complete and reciprocal PC matrix, experts need to perform $n(n-2)$ comparisons. Where the number of alternatives is large, this can be a tedious and difficult task[1]. *Harker* [76] points out that experts may prefer not to provide direct answers to certain comparisons. This can happen when the comparison concerns two sensitive criteria, e.g., mortality risk vs. cost, and the experts do not want to present in public their statement on this tradeoff. In addition, experts may not be completely sure of their preferences as to the two compared alternatives. In this situation, they may prefer not to indicate the winner in a direct comparison. In all these cases, a good solution is not to force experts to evaluate two problematic alternatives but accept an incomplete PC matrix (Section 2.3) as the basis for calculating the ranking. Selected priority-deriving methods for incomplete PC matrices are presented below.

---

[1]The desire to reduce the number of comparisons was one of the reasons for the BWM method [151, 124].

4.1   Eigenvalue method for incomplete PC matrices

4.1.1   Idea of the method

An extension of the EV method (Sec. 3.1) for incomplete PC matrices (Sec. 2.3) has been proposed by *Harker* [76]. He noticed that since $c_{ij} \approx w(a_i)/w(a_j)$ for every $i, j = 1, \ldots, n$ (3.2, 3.3) then this approximation also holds for the missing values. Hence, he proposed the replacement of the missing values $c_{ij} =?$ with fractions in the form $w(a_i)/w(a_j)$, where $w(a_i)$ and $w(a_j)$ are the values of the final ranking[2]. Thus, instead of considering matrices like

$$C = \begin{pmatrix} 1 & ? & ? & c_{14} \\ ? & 1 & c_{23} & ? \\ ? & 1/c_{23} & 1 & c_{34} \\ 1/c_{14} & ? & 1/c_{34} & 1 \end{pmatrix} \tag{4.1}$$

we try to find the ranking of

$$C^* = \begin{pmatrix} 1 & \frac{w(a_1)}{w(a_2)} & \frac{w(a_1)}{w(a_3)} & c_{14} \\ \frac{w(a_2)}{w(a_1)} & 1 & c_{23} & \frac{w(a_2)}{w(a_4)} \\ \frac{w(a_3)}{w(a_1)} & c_{32} & 1 & c_{34} \\ c_{41} & \frac{w(a_4)}{w(a_2)} & c_{43} & 1 \end{pmatrix}. \tag{4.2}$$

Of course, the ranking vector $w = [w(a_1), \ldots, w(a_4)]^T$ is a priori unknown, hence C^* contains both: real numbers c_{ij} and variables $w(a_i)/w(a_j)$. Therefore, we cannot compute the principal eigenvector for C^* directly. We can, however, consider the matrix equation

$$C^* w = \lambda_{max} w,$$

where λ_{max} is the principal eigenvalue and w is the principal eigenvector of C^*. This equation translates into the following equation system

$$\begin{aligned}
w(a_1) + \frac{w(a_1)}{w(a_2)} w(a_2) + \frac{w(a_1)}{w(a_3)} w(a_3) + c_{14} w(a_4) &= \lambda_{max} \cdot w(a_1) \\
\frac{w(a_2)}{w(a_1)} w(a_1) + w(a_2) + c_{23} w(a_3) + \frac{w(a_2)}{w(a_4)} w(a_4) &= \lambda_{max} \cdot w(a_2) \\
\frac{w(a_3)}{w(a_1)} w(a_1) + c_{32} w(a_2) + w(a_3) + c_{34} w(a_4) &= \lambda_{max} \cdot w(a_3) \\
c_{41} w(a_1) + \frac{w(a_4)}{w(a_2)} w(a_2) + c_{43} w(a_3) + w(a_4) &= \lambda_{max} \cdot w(a_4)
\end{aligned}$$

Thus,

$$\begin{aligned}
w(a_1) + w(a_1) + w(a_1) + c_{14} w(a_4) &= \lambda_{max} \cdot w(a_1) \\
w(a_2) + w(a_2) + c_{23} w(a_3) + w(a_2) &= \lambda_{max} \cdot w(a_2) \\
w(a_3) + c_{32} w(a_2) + w(a_3) + c_{34} w(a_4) &= \lambda_{max} \cdot w(a_3) \\
c_{41} w(a_1) + w(a_4) + c_{43} w(a_3) + w(a_4) &= \lambda_{max} \cdot w(a_4)
\end{aligned}$$

---

[2]Note that here w stands for the principal eigenvector.

which may be written as

$$
\begin{array}{rcl}
3w(a_1) + 0 + 0 + c_{14}w(a_4) & = & \lambda_{max} \cdot w(a_1) \\
0 + 3w(a_2) + c_{23}w(a_3) + 0 & = & \lambda_{max} \cdot w(a_2) \\
0 + c_{32}w(a_2) + 2w(a_3) + c_{34}w(a_4) & = & \lambda_{max} \cdot w(a_3) \\
c_{41}w(a_1) + 0 + c_{43}w(a_3) + 2w(a_4) & = & \lambda_{max} \cdot w(a_4)
\end{array}
$$

The matrix form of the above equation system is as follows:

$$
\begin{pmatrix}
3 & 0 & 0 & c_{14} \\
0 & 3 & c_{23} & 0 \\
0 & c_{32} & 2 & c_{34} \\
c_{41} & 0 & c_{43} & 2
\end{pmatrix}
\begin{pmatrix}
w(a_1) \\
w(a_2) \\
w(a_3) \\
w(a_4)
\end{pmatrix}
= \lambda_{max}
\begin{pmatrix}
w(a_1) \\
w(a_2) \\
w(a_3) \\
w(a_4)
\end{pmatrix}.
$$

Hence, the new matrix equation takes the form

$$
Bw = \lambda_{max}w, \tag{4.3}
$$

where

$$
B =
\begin{pmatrix}
3 & 0 & 0 & c_{14} \\
0 & 3 & c_{23} & 0 \\
0 & c_{32} & 2 & c_{34} \\
c_{41} & 0 & c_{43} & 2
\end{pmatrix}.
$$

In this way, we get rid of variables in the form of $w(a_i)/w(a_j)$ from the matrix B, thus we can easily solve (4.3) and, as follows, we can easily calculate the principal eigenvector w. In other words, to calculate the ranking for C we prepared the auxiliary matrix B, whose principal eigenvector has been accepted as the result of the ranking based on C. This reasoning can be generalized to any incomplete matrix.

4.1.2 Ranking algorithm

The following algorithm emerges from the above considerations:

1. Based on the given incomplete PC matrix C, let us prepare the auxiliary matrix $B = [b_{ij}]$ such that:

$$
b_{ij} =
\begin{cases}
0 & \text{if } c_{ij} =? \text{ and } i \neq j \\
c_{ij} & \text{if } c_{ij} \neq? \text{ and } i \neq j\,, \\
s_i + 1 & \text{if } i = j
\end{cases}
\tag{4.4}
$$

where s_i is the number of missing comparisons in the i-th row of C.

2. Find the principal eigenvector w_{max} of B such that

$$
Bw_{max} = \lambda_{max}w_{max}.
$$

3. Rescale w_{max} so that all its entries sum up to 1, and adopt the resulting vector as the result of the ranking

$$
w_{ev} = \begin{pmatrix} \frac{w_{max}(a_1)}{\sum_{j=1}^{n} w_{max}(a_j)} \\ \frac{w_{max}(a_2)}{\sum_{j=1}^{n} w_{max}(a_j)} \\ \vdots \\ \frac{w_{max}(a_n)}{\sum_{j=1}^{n} w_{max}(a_j)} \end{pmatrix}.
$$

It is worth noting that if C is complete then $B = C$, hence Harker's extension presented above boils down to EVM as shown in (Sec. 3.1). It should also be remembered that the presented method is applicable if and only if the incomplete matrix is irreducible (Def. 7). More about the extension of EVM for incomplete PC matrices can be found in [76].

4.1.3   Illustrative example

Example 5. *Let us consider a 5×5 incomplete PC matrix corresponding to the pairwise judgments of five alternatives $a_1, \ldots, a_5$:*

$$
C = \begin{pmatrix} 1 & \frac{2}{3} & ? & ? & 9 \\ \frac{3}{2} & 1 & ? & \frac{7}{4} & ? \\ ? & ? & 1 & ? & \frac{1}{3} \\ ? & \frac{4}{7} & ? & 1 & 9 \\ \frac{1}{9} & ? & 3 & \frac{1}{9} & 1 \end{pmatrix}.
$$

The auxiliary Harker matrix B built based on C is

$$
B = \begin{pmatrix} 3 & \frac{2}{3} & 0 & 0 & 9 \\ \frac{3}{2} & 3 & 0 & \frac{7}{4} & 0 \\ 0 & 0 & 4 & 0 & \frac{1}{3} \\ 0 & \frac{4}{7} & 0 & 3 & 9 \\ \frac{1}{9} & 0 & 3 & \frac{1}{9} & 2 \end{pmatrix}.
$$

hence, the rescaled principal eigenvector of B is given as

$$
w_{ev} = \begin{pmatrix} 0.275 \\ 0.429 \\ 0.0098 \\ 0.255 \\ 0.0294 \end{pmatrix},
$$

and the principal eigenvalue is $\lambda_{max} = 5.00119$. Thus, according to the created ranking, the most important (the most preferred) alternative is a_2 with the priority $w(a_2) = 0.429$. The second, third, fourth and fifth places are taken by a_1, a_4, a_5 and a_3, correspondingly.

It is easy to verify that indeed

$$
\begin{pmatrix}
1 & \frac{2}{3} & \frac{0.275}{0.0098} & \frac{0.275}{0.255} & 9 \\
\frac{3}{2} & 1 & \frac{0.429}{0.0098} & \frac{7}{4} & \frac{0.429}{0.029} \\
\frac{0.0098}{0.275} & \frac{0.0098}{0.429} & 1 & \frac{0.0098}{0.255} & \frac{1}{3} \\
\frac{0.255}{0.275} & \frac{4}{7} & \frac{0.255}{0.0098} & 1 & 9 \\
\frac{1}{9} & \frac{0.0294}{0.429} & 3 & \frac{1}{9} & 1
\end{pmatrix}
\begin{pmatrix}
0.275 \\
0.429 \\
0.0098 \\
0.255 \\
0.029
\end{pmatrix}
= 5.001
\begin{pmatrix}
0.275 \\
0.429 \\
0.0098 \\
0.255 \\
0.029
\end{pmatrix},
$$

which is in line with the assumptions of the solution proposed by Harker.

4.2 Geometric mean method for incomplete PC matrices

4.2.1 Idea of the method

The GMM (Sec. 3.2) also has its own counterpart for incomplete PC matrices. This extension has been proposed in [114]. Some similar ideas can also be found in [1]. In general, as before, the idea is based on replacing in C the missing entries $c_{ij} = ?$ by the ratios $w(a_i)/w(a_j)$, composed of the elements of the final ranking. Thus, instead of considering incomplete matrices such as C (4.1), we consider the matrices similar to C^* (4.2) composed of real numbers and variables. As before, when starting from the postulate

$$
w(a_i) = \sqrt[n]{\prod_{j=1}^{n} c_{ij} w(a_j)} \quad \text{for} \quad i = 1, \ldots, n
$$

it is easy to show (see Sec. 3.2) that for the purpose of comparisons of ranking values it is enough to adopt

$$
w(a_i) = \sqrt[n]{\prod_{j=1}^{n} c_{ij}} \quad \text{for} \quad i = 1, \ldots, n.
$$

However, in the case of C^* (4.2) this leads to the following non-linear equation system:

$$
\begin{aligned}
\left(1 \cdot \frac{w(a_1)}{w(a_2)} \cdot \frac{w(a_1)}{w(a_3)} \cdot c_{14} \right)^{\frac{1}{4}} &= w(a_1) \\
\left(\frac{w(a_2)}{w(a_1)} \cdot 1 \cdot c_{23} \cdot \frac{w(a_2)}{w(a_4)} \right)^{\frac{1}{4}} &= w(a_2) \\
\left(\frac{w(a_3)}{w(a_1)} \cdot c_{32} \cdot 1 \cdot c_{34} \right)^{\frac{1}{4}} &= w(a_3) \\
\left(c_{41} \cdot \frac{w(a_4)}{w(a_2)} \cdot c_{43} \cdot 1 \right)^{\frac{1}{4}} &= w(a_4)
\end{aligned}
.
$$

To solve it, let us raise both sides to the power,

$$
\begin{aligned}
1 \cdot \frac{w(a_1)}{w(a_2)} \cdot \frac{w(a_1)}{w(a_3)} \cdot c_{14} &= w^4(a_1) \\
\frac{w(a_2)}{w(a_1)} \cdot 1 \cdot c_{23} \cdot \frac{w(a_2)}{w(a_4)} &= w^4(a_2) \\
\frac{w(a_3)}{w(a_1)} \cdot c_{32} \cdot 1 \cdot c_{34} &= w^4(a_3) \\
c_{41} \cdot \frac{w(a_4)}{w(a_2)} \cdot c_{43} \cdot 1 &= w^4(a_4)
\end{aligned}
$$

and apply logarithm transformation to both sides. Then, we obtain

$$
\begin{aligned}
\ln \left(1 \cdot \frac{1}{w(a_2)} \cdot \frac{1}{w(a_3)} \cdot c_{14} \right) &= 2\ln w(a_1) \\
\ln \left(\frac{1}{w(a_1)} \cdot 1 \cdot c_{23} \cdot \frac{1}{w(a_4)} \right) &= 2\ln w(a_2) \\
\ln \left(\frac{1}{w(a_1)} \cdot c_{32} \cdot 1 \cdot c_{34} \right) &= 3\ln w(a_3) \\
\ln \left(c_{41} \cdot \frac{1}{w(a_2)} \cdot c_{43} \cdot 1 \right) &= 3\ln w(a_4)
\end{aligned}
$$

and

$$
\begin{aligned}
0 - \ln w(a_2) - \ln w(a_3) + \ln c_{14} &= 2\ln w(a_1) \\
- \ln w(a_1) + 0 + \ln c_{23} - \ln w(a_4) &= 2\ln w(a_2) \\
- \ln w(a_1) + \ln c_{32} + 0 + \ln c_{34} &= 3\ln w(a_3) \\
\ln c_{41} - \ln w(a_2) + \ln c_{43} + 0 &= 3\ln w(a_4)
\end{aligned} \tag{4.5}
$$

The above equations (4.5) can be written as the following linear equation system

$$
\begin{aligned}
2\ln w(a_1) + \ln w(a_2) + \ln w(a_3) + 0 &= \ln c_{14} \\
\ln w(a_1) + 2\ln w(a_2) + 0 + \ln w(a_4) &= \ln c_{23} \\
\ln w(a_1) + 0 + 3\ln w(a_3) + 0 &= +\ln c_{32} + \ln c_{34} \\
0 + \ln w(a_2) + 0 + 3\ln w(a_4) &= +\ln c_{41} + \ln c_{43}
\end{aligned}
$$

Let us denote $\log w(a_i) \overset{df}{=} \widehat{w}(a_i)$ and $\ln c_{ij} \overset{df}{=} \widehat{c}_{ij}$. Hence, the above equation system can be written down in the form of a matrix

$$
B\widehat{w} = \widehat{c}, \tag{4.6}
$$

where the auxiliary matrix B is given as

$$
B = \begin{pmatrix} 2 & 1 & 1 & 0 \\ 1 & 2 & 0 & 1 \\ 1 & 0 & 3 & 0 \\ 0 & 1 & 0 & 3 \end{pmatrix},
$$

where the constant term vector $\widehat{c}$ and the vector $\widehat{w}$ are as follows:

$$
\widehat{c} = \begin{pmatrix} \widehat{c}_{14} \\ \widehat{c}_{23} \\ \widehat{c}_{32} + \widehat{c}_{34} \\ \widehat{c}_{41} + \widehat{c}_{43} \end{pmatrix}, \quad \widehat{w} = \begin{pmatrix} \widehat{w}(a_1) \\ \widehat{w}(a_2) \\ \widehat{w}(a_3) \\ \widehat{w}(a_4) \end{pmatrix}.
$$

The matrix B is non-singular, thus we may compute the auxiliary vector:

$$\hat{w} = \begin{pmatrix} \frac{15\hat{c}_{14}-9\hat{c}_{23}-5\hat{c}_{32}-5\hat{c}_{34}+3\hat{c}_{41}+3\hat{c}_{43}}{16} \\ \frac{-9\hat{c}_{14}+15\hat{c}_{23}+3\hat{c}_{32}+3\hat{c}_{34}-5\hat{c}_{41}-5\hat{c}_{43}}{16} \\ \frac{-5\hat{c}_{14}+3\hat{c}_{23}+7\hat{c}_{32}+7\hat{c}_{34}-\hat{c}_{41}-\hat{c}_{43}}{16} \\ \frac{3\hat{c}_{14}-5\hat{c}_{23}-\hat{c}_{32}-\hat{c}_{34}+7\hat{c}_{41}+7\hat{c}_{43}}{16} \end{pmatrix}.$$

Finally, by exponential transformation, we receive the unscaled ranking vector:

$$w = \begin{pmatrix} e^{\frac{15\hat{c}_{14}-9\hat{c}_{23}-5\hat{c}_{32}-5\hat{c}_{34}+3\hat{c}_{41}+3\hat{c}_{43}}{16}} \\ e^{\frac{-9\hat{c}_{14}+15\hat{c}_{23}+3\hat{c}_{32}+3\hat{c}_{34}-5\hat{c}_{41}-5\hat{c}_{43}}{16}} \\ e^{\frac{-5\hat{c}_{14}+3\hat{c}_{23}+7\hat{c}_{32}+7\hat{c}_{34}-\hat{c}_{41}-\hat{c}_{43}}{16}} \\ e^{\frac{3\hat{c}_{14}-5\hat{c}_{23}-\hat{c}_{32}-\hat{c}_{34}+7\hat{c}_{41}+7\hat{c}_{43}}{16}} \end{pmatrix}.$$

After scaling, we obtain

$$w_{gm} = \frac{1}{S}w, \tag{4.7}$$

where

$$S = e^{\frac{15\hat{c}_{14}-9\hat{c}_{23}-5\hat{c}_{32}-5\hat{c}_{34}+3\hat{c}_{41}+3\hat{c}_{43}}{16}} + e^{\frac{-9\hat{c}_{14}+15\hat{c}_{23}+3\hat{c}_{32}+3\hat{c}_{34}-5\hat{c}_{41}-5\hat{c}_{43}}{16}} +$$
$$+ e^{\frac{-5\hat{c}_{14}+3\hat{c}_{23}+7\hat{c}_{32}+7\hat{c}_{34}-\hat{c}_{41}-\hat{c}_{43}}{16}} + e^{\frac{3\hat{c}_{14}-5\hat{c}_{23}-\hat{c}_{32}-\hat{c}_{34}+7\hat{c}_{41}+7\hat{c}_{43}}{16}}.$$

Based on (4.7) and knowing the actual values of c_{14}, c_{23} and c_{34}, we are able to calculate the final numerical ranking.

4.2.2 Ranking algorithm

The presented reasoning leads to the following algorithm:

1. Based on the given incomplete PC matrix C, let us create the auxiliary matrix $G = [g_{ij}]$, such that

$$g_{ij} = \begin{cases} 1 & \text{if } c_{ij} =? \text{ and } i \neq j \\ 0 & \text{if } c_{ij} \neq ? \text{ and } i \neq j \,, \\ n - s_i & \text{if } i = j \end{cases}$$

where s_i is the number of missing comparisons in the i-th row of C, and the constant term vector

$$r = \begin{pmatrix} \sum_{\substack{j=1 \\ c_{1j}\neq?}}^{n} \ln c_{1j} \\ \vdots \\ \vdots \\ \sum_{\substack{j=1 \\ c_{nj}\neq?}}^{n} \ln c_{nj} \end{pmatrix}. \tag{4.8}$$

2. Solve the following linear equation system

$$G\widehat{w} = r \qquad (4.9)$$

and create the ranking vector w in the form:

$$w = \begin{pmatrix} e^{\widehat{w}(a_1)} \\ \vdots \\ \vdots \\ e^{\widehat{w}(a_n)} \end{pmatrix}.$$

3. Rescale w so that all its entries sum up to 1, and obtain the resulting vector as the result of the ranking

$$w_{gm} = \begin{pmatrix} \dfrac{w(a_1)}{\sum_{j=1}^{n} w(a_j)} \\ \dfrac{w(a_2)}{\sum_{j=1}^{n} w(a_j)} \\ \vdots \\ \dfrac{w(a_n)}{\sum_{j=1}^{n} w(a_j)} \end{pmatrix}.$$

As before, if C is complete then the above method boils down to GMM (the only difference is that instead of computing geometric mean we apply the logarithmic transformation). The above method can be used if and only if the incomplete matrix is irreducible (Def. 7), i.e., the graph of C is connected. As will be shown in (Section 4.2.4), connectivity is a condition for the existence of a solution of (4.9).

4.2.3   Illustrative example

Example 6. *Let us consider a 5×5 incomplete PC matrix corresponding to the pairwise judgments of five alternatives $a_1, \ldots, a_5$:*

$$C = \begin{pmatrix} 1 & \frac{2}{3} & ? & ? & 9 \\ \frac{3}{2} & 1 & ? & \frac{7}{4} & ? \\ ? & ? & 1 & ? & \frac{1}{3} \\ ? & \frac{4}{7} & ? & 1 & 9 \\ \frac{1}{9} & ? & 3 & \frac{1}{9} & 1 \end{pmatrix}.$$

The auxiliary matrix G is as follows:

$$G = \begin{pmatrix} 3 & 0 & 1 & 1 & 0 \\ 0 & 3 & 1 & 0 & 1 \\ 1 & 1 & 2 & 1 & 0 \\ 1 & 0 & 1 & 3 & 0 \\ 0 & 1 & 0 & 0 & 4 \end{pmatrix},$$

and the constant term vector r is given as

$$r = \begin{pmatrix} \ln \frac{9}{2} \\ 0 \\ -\ln 9 \\ -\ln 54 \\ -\ln 108 \end{pmatrix}.$$

Solving $G\widehat{w} = r$ leads to the following logarithmized ranking vector

$$\widehat{w} = \begin{pmatrix} \frac{1}{100}\left(22\ln 9 + 7\ln 54 - 2\ln 108 + 43\ln 9/2\right) \\ \frac{1}{25}\left(8\ln 9 - 2\ln 54 - 3\ln 108 + 2\ln 9/2\right) \\ \frac{1}{50}\left(-44\ln 9 + 11\ln 54 + 4\ln 108 - 11\ln 9/2\right) \\ \frac{1}{100}\left(22\ln 9 - 43\ln 54 - 2\ln 108 - 7\ln 9/2\right) \\ \frac{1}{50}\left(-4\ln 9 + \ln 54 + 14\ln 108 - \ln 9/2\right) \end{pmatrix} = \begin{pmatrix} 1.3157 \\ -0.0575 \\ -1.0123 \\ -1.4308 \\ 1.1849 \end{pmatrix}.$$

Hence, the (unscaled) ranking vector is

$$w = \begin{pmatrix} e^{1.3157} \\ e^{-0.0575} \\ e^{-1.0123} \\ e^{-1.4308} \\ e^{1.1849} \end{pmatrix} = \begin{pmatrix} 3.727 \\ 0.944 \\ 0.363 \\ 0.239 \\ 3.27 \end{pmatrix}.$$

The last step to receive the ranking in the usual form is scaling so that the entries of the ranking vector sum up to 1. The final form of the ranking vector is as follows:

$$w_{gm} = \begin{pmatrix} 0.436 \\ 0.11 \\ 0.042 \\ 0.027 \\ 0.382 \end{pmatrix}.$$

According to the computed ranking, the most preferred alternative is a_1 with the ranking value $w(a_1) = 0.436$. The second place is taken by a_5 with $w(a_5) = 0.382$, then a_2, a_3 and a_4.

Of course, one may verify that GMM applied to the following matrix

$$\begin{pmatrix} 1 & \frac{2}{3} & \frac{0.436}{0.042} & \frac{0.436}{0.027} & 9 \\ \frac{3}{2} & 1 & \frac{0.11}{0.042} & \frac{7}{4} & \frac{0.11}{0.382} \\ \frac{0.042}{0.436} & \frac{0.042}{0.11} & 1 & \frac{0.042}{0.027} & \frac{1}{3} \\ \frac{0.027}{0.436} & \frac{4}{7} & \frac{0.027}{0.042} & 1 & 9 \\ \frac{1}{9} & \frac{0.382}{0.11} & 3 & \frac{1}{9} & 1 \end{pmatrix}$$

results in w_{gm}.

4.2.4   Solution existence

It is easy to see that the above procedure makes sense only if (4.9) has a solution. Since, due to (4.9) $\widehat{w} = G^{-1}r$, then it is enough to show that G is invertible and real. In fact, following [114], we show that G is an M-matrix, hence due to the *Plemmons'* characterization [143] G is inverse positive, i.e., G^{-1} exists and $G^{-1} \geq 0$. Let us note that G can be written in the form:

$$G = L(P_C) + J_n,$$

where $L(P_C)$ is the Laplacian matrix (Section 2.5) of a graph P_C, and J_n is the n by n matrix with each entry equal to 1.

Theorem 2. *The matrix $G = L(G_C) + J_n$ is an M-matrix*

Proof. Let $\lambda_1, \ldots, \lambda_n$ be the eigenvalues of $L(P_C) = [l_{ij}]$ such that $\lambda_1 \geq \ldots \geq \lambda_n$ and $w_1, \ldots, w_n$ are their eigenvectors. Due to the fact that $L(P_C)$ is a Laplacian matrix, all its eigenvalues are non-negative and real. Moreover, the irreducibility of C implies that $\lambda_{n-1} > 0$ [128, p. 147]. Since for every row of $L(P_C)$ it holds that $l_{ii} = \sum_{j=1, i \neq j}^{n} |l_{ij}|$, then we may deduce that

$$L(P_C) \begin{pmatrix} 1 \\ 1 \\ \vdots \\ 1 \end{pmatrix} = 0,$$

i.e., $w_n = (1, 1, \ldots, 1)^T$. As $L(P_C)$ is real and symmetric, then the eigenvectors corresponding to different eigenvalues are orthogonal . In particular, all eigenvectors $w_1, \ldots, w_{n-1}$ are orthogonal to w_n, i.e., $w_n^T w_i = 0$ for $i = 1, \ldots, n$.

Let us consider the equation:

$$Gw_i = \widehat{\lambda}_i w_i \quad \text{for } i = 1, \ldots, n.$$

This can be written in the form:

$$(L(P_C) + J_n) w_i = \widehat{\lambda}_i w_i$$

i.e.,

$$L(P_C)w_i + J_n w_i = \widehat{\lambda}_i w_i.$$

When $i = n$ then $L(P_C)w_i = 0$ and

$$J_n w_n = n w_n,$$

thus,

$$G w_n = n w_n,$$

hence $\widehat{\lambda}_n = n$. For $i = 1, \ldots, n-1$ we have $L(P_C)w_i = \lambda_i w_i$ and

$$
J_n w_i = \begin{pmatrix} w_n^T w_i \\ w_n^T w_i \\ \vdots \\ w_n^T w_i \end{pmatrix} = \begin{pmatrix} 0 \\ 0 \\ \vdots \\ 0 \end{pmatrix}.
$$

Thus,

$$
\widehat{\lambda}_i w_i = G w_i = \left(L(P_C) + J_n \right) w_i = L(P_C) w_i = \lambda_i w_i.
$$

In other words, G inherits from $L(P_C)$ the first $n - 1$ eigenvalues $\lambda_1 = \widehat{\lambda}_1, \ldots, \lambda_{n-1} = \widehat{\lambda}_{n-1}$ while the n-th eigenvalue $\lambda_n = 0$ changes to $\widehat{\lambda}_n = n$. Since all the eigenvalues G are real and positive then, due to *Plemmons'* theorem [143], G is an M-matrix, which implies that G is inverse positive, i.e., G^{-1} exists and $G^{-1} \geq 0$. Thus, $G\widehat{w} = r$ has a unique solution. $\qquad\square$

4.2.5   Optimality

As we know, the geometric mean method for complete PC matrices is considered as optimal (Section 3.2) because it minimizes the total logarithmic error $\mathcal{E}$ for the pairwise comparisons matrix C. In other words,

$$
\mathcal{E}(C, w) = \sum_{i,j=1}^{n} \left(\ln c_{ij} - \ln \frac{w(a_i)}{w(a_j)} \right)^2
$$

is going to be minimal when C is a PC matrix and w is the priority vector obtained from GMM. For an incomplete PC matrix, however, some c_{ij} are undefined. It is natural to exclude them from the above optimality condition. Hence, the more general formula [23, 25] looks as follows:

$$
\mathcal{E}^*(C, w) = \sum_{\substack{i,j=1 \\ c_{ij} \neq ?}}^{n} \left(\ln c_{ij} - \ln \frac{w(a_i)}{w(a_j)} \right)^2. \tag{4.10}
$$

The generalized formula allows us to consider the problem of optimality in the context of incomplete matrices. Taking advantage of the fact that GMM is optimal (Section 3.2.3), we are able to prove that the extended GMM for incomplete PC matrices (Section 4.2.2) is also optimal. To do this, it is enough to demonstrate that w_{gm} minimizes $\mathcal{E}^*(C, w_{gm})$ for every irreducible PC matrix C.

Theorem 3. *The total logarithmic error $\mathcal{E}^*(C, w_{gm})$ is minimal if C is an incomplete and irreducible PC matrix and w_{gm} is the result of GMM for an incomplete PC matrix.*

Proof. Since C is irreducible then (4.9) has a solution (Section 4.2.4) and, as follows, w_{gm} exists such that

$$w_{gm}(a_i) = \sqrt[n]{\prod_{j=1}^{n} c_{ij}^*},$$

and $C^* = [c_{ij}^*]$, where

$$c_{ij}^* = \begin{cases} c_{ij} & \text{if } c_{ij} \neq ? \\ \dfrac{w_{gm}(a_i)}{w_{gm}(a_j)} & \text{if } c_{ij} = ? \end{cases}. \tag{4.11}$$

Since GMM is optimal (Section 3.2.3), then

$$\mathcal{E}(C^*, w_{gm}) = \sum_{i,j=1}^{n} \left(\ln c_{ij}^* - \ln \frac{w(a_i)}{w(a_j)} \right)^2$$

is minimal. The above formula can be written as the sum of two components:

$$\mathcal{E}(C^*, w_{gm}) = \sum_{\substack{i,j=1 \\ c_{ij} \neq ?}}^{n} \left(\ln c_{ij}^* - \ln \frac{w_{gm}(a_i)}{w_{gm}(a_j)} \right)^2 + \sum_{\substack{i,j=1 \\ c_{ij} = ?}}^{n} \left(\ln c_{ij}^* - \ln \frac{w_{gm}(a_i)}{w_{gm}(a_j)} \right)^2.$$

Hence, due to the form of the matrix C^* (4.11)

$$\mathcal{E}(C^*, w_{gm}) = \sum_{\substack{i,j=1 \\ c_{ij} \neq ?}}^{n} \left(\ln c_{ij} - \ln \frac{w_{gm}(a_i)}{w_{gm}(a_j)} \right)^2 + \sum_{\substack{i,j=1 \\ c_{ij} = ?}}^{n} \left(\ln \frac{w_{gm}(a_i)}{w_{gm}(a_j)} - \ln \frac{w_{gm}(a_i)}{w_{gm}(a_j)} \right)^2,$$

i.e.,

$$\mathcal{E}(C^*, w_{gm}) = \sum_{\substack{i,j=1 \\ c_{ij} \neq ?}}^{n} \left(\ln c_{ij} - \ln \frac{w_{gm}(a_i)}{w_{gm}(a_j)} \right)^2$$

is minimal. The above equation, however, is equivalent to

$$\mathcal{E}^*(C, w_{gm}) = \sum_{\substack{i,j=1 \\ c_{ij} \neq ?}}^{n} \left(\ln c_{ij} - \ln \frac{w_{gm}(a_i)}{w_{gm}(a_j)} \right)^2,$$

which means that $\mathcal{E}^*(C, w_{gm})$ is minimal.   □

4.3 Logarithmic least square method

4.3.1 Idea of the method

The Logarithmic Least Square Method (LLSM) for incomplete PC matrices has been proposed by *Tone* [191] and *Bozóki, Fülöp* and *Rónyaj* [23]. It is an optimization method which is based (as LLSM for complete matrices (Section 3.3.3)) on the observation that $w(a_i)/w(a_j)$ approximates the value c_{ij} in the given PC matrix C. The minimization criterion takes the form of a sum of logarithm differences, however, only the components for which $c_{ij} \neq ?$ are taken into account (4.10). Hence, the Logarithmic Least Square (LLS) problem [121, 189, 25] is formulated as

$$\min \quad \mathcal{E}^*(C, w) = \sum_{\substack{i,j=1 \\ c_{ij} \neq ?}}^{n} \left(\ln c_{ij} - \ln \frac{w(a_i)}{w(a_j)} \right)^2$$

$$\text{s.t.} \quad w(a_i) > 0 \text{ for } i = 1, \ldots, n$$

The optimal solution of the above optimization problem can be obtained by solving the following linear equation system

$$L(P_C)\widehat{w} = r$$

with the constraint

$$\widehat{w}(a_1) = 0, \tag{4.12}$$

where $L(P_C)$ is the Laplacian matrix corresponding to the graph of C, $\widehat{w} = e^y$, and the constant term vector r (4.8) [23, 25]. The constraint (4.12) has a technical meaning. For example, instead of it, one may assume that $\sum_{i=1}^{n} \widehat{w}(a_i) = 0$ which would be equivalent to $\prod_{i=1}^{n} \widehat{w}(a_i) = 0$ [25].

Both GMM and LLSM methods for incomplete PC matrices are equivalent and lead to the same ranking vectors. This is a natural consequence of the fact that both methods minimize $\mathcal{E}^*(C, w)$.

4.3.2 Ranking algorithm

The algorithm allowing the LLSM ranking vector to be calculated is as follows:

1. Create the auxiliary (Laplacian) matrix $L = [l_{ij}]$, such that

$$l_{ij} = \begin{cases} 0 & \text{if } c_{ij} =? \text{ and } i \neq j \\ -1 & \text{if } c_{ij} \neq ? \text{ and } i \neq j, \\ p_i & \text{if } i = j \end{cases}$$

where p_i is the number of existing comparisons in the i-th row of C, and the constant term vector

$$
r = \begin{pmatrix} \sum_{\substack{j=1 \\ c_{1j} \neq ?}}^{n} \ln c_{1j} \\ \vdots \\ \vdots \\ \sum_{\substack{j=1 \\ c_{nj} \neq ?}}^{n} \ln c_{nj} \end{pmatrix}.
$$

2. Solve the equation system

$$
L\widehat{w} = r
$$
$$
\widehat{w}(a_1) = 0,
$$

3. Compute the ranking vector as

$$
w = \begin{pmatrix} e^{\widehat{w}(a_1)} \\ \vdots \\ \vdots \\ e^{\widehat{w}(a_n)} \end{pmatrix}.
$$

4. If needed, rescale the vector w so that all its entries sum up to 1.

4.4   Optimization methods

In addition to LLSM, other optimization methods discussed in (Section 3.3) can also be used to determine the priority vector based on an incomplete PC matrix. Of course, in this case, the optimization criteria must be narrowed down to the existing elements in the matrix.

4.4.1   Weighted least absolute error method

For example, *Takeda* and *Yu* [189] propose the solution of a slightly modified problem, which is the optimization criterion in the weighted least absolute error method (Section 3.3.6). Hence, the priorities $w(a_1), \ldots w(a_n)$ are the results of solving the following linear program:

$$
\min \quad \sum_{\substack{i,j=1 \\ c_{ij} \neq ?}}^{n} \varepsilon_{ij},
$$

subject to

$$|c_{ij}w(a_j) - w(a_i)| \leq \varepsilon_{i,j} \text{ and } c_{ij} \neq ?,$$

$$\sum_{i=1}^{n} w(a_i) = 1$$

and

$$w(a_i) > 0 \text{ for } i = 1, \ldots, n.$$

4.4.2 Inconsistency minimization

Although various inconsistency indices are discussed later in Chapter 6, it is worth noting yet another strategy that may allow us to deal with the problem of rankings for incomplete PC matrices. This strategy is to find such a completion of a PC matrix that the inconsistency measured by the given index is minimal. An example of this approach can be q_3 coefficient maximization [27]. *Shiraishi et al.* [177] noticed that for a characteristic polynomial of C given as

$$P(C) = x^n + q_1 x^{n-1} + \ldots + q_{n-1}x + q_n$$

its third coefficient q_3 is well suited for measuring inconsistency. Moreover, q_3 is given by the following analytic formula

$$q_3 = \sum_{i=1}^{n-2} \sum_{j=i+1}^{n-1} \sum_{k=j+1}^{n} \left(2 - \frac{a_{ik}}{a_{ij}a_{jk}} - \frac{a_{ij}a_{jk}}{a_{ik}} \right).$$

The higher the value of q_3, the lower the inconsistency of C. Hence, for the incomplete PC matrix $C = [c_{ij}]$ *Shiraishi* et al. propose the solution of the following optimization problem:

$$\max_{\substack{c_{ij}=? \\ i,j=1,\ldots,n}} q_3$$

subject to

$$c_{ij} > 0 \text{ for } i, j = 1, \ldots, n.$$

It is proven that the above problem always has a solution, although when there are too many missing comparisons the uniqueness may not be preserved [27, 176].

When the missing values are determined, a priority ranking vector can be computed using standard EVM (Section 3.1).

4.5 Other optimization methods

4.5.1 Null entries approximation

The strategy proposed by *Shiraishi* et al. [177] consists of finding missing values in some way (here by solving the optimization problem), then calculating the ranking for such a completed matrix. An example of using a similar strategy is completing null entries proposed by *Koczkodaj et al.* [100]. The authors start from the observation that for every PC matrix C it holds that

$$c_{ij} \approx c_{ik}c_{kj} \text{ for } i,j,k = 1,\ldots,n.$$

This relationship can be extended to all products in the form $c_{ij}c_{kj}$, i.e.,

$$c_{ij}^n \approx \prod_{k=1}^{n} c_{ik}c_{kj}.$$

The above formula might be used to create an approximation of c_{ij} as the geometric mean of $c_{ij}c_{kj}$ products. Hence,

$$c_{ij} \approx \left(\prod_{k=1}^{n} c_{ik}c_{kj} \right)^{\frac{1}{n}},$$

which can be written as

$$c_{ij} \approx \left(\prod_{k=1}^{n} c_{ik} \right)^{\frac{1}{n}} \left(\prod_{k=1}^{n} c_{kj} \right)^{\frac{1}{n}}.$$

This approximation is also valid for incomplete PC matrices to the extent that the missing values need to be skipped in the above formula. This leads to the proposal made in [100]:

$$c_{ij} \approx \left(\prod_{\substack{k=1 \\ c_{ik} \neq ?}}^{n} c_{ik} \right)^{\frac{1}{x}} \left(\prod_{\substack{k=1 \\ c_{kj} \neq ?}}^{n} c_{kj} \right)^{\frac{1}{y}}, \tag{4.13}$$

where x means the number of missing values in the i-th row and y is the number of missing entries in the j-th column of C. It is worth noting that c_{ij} can also be a missing value. This approach does not explicitly assume that C must be irreducible. However, the higher the completeness of C the better. Otherwise, it is possible that (in an extreme situation) in the particular row and/or column there will be no entries except the diagonal. In such a case, it is difficult to say that (4.13) is a good approximation of a missing value. Nevertheless, in the situation in which we do not have a tool with which we can solve the linear equations system (Section 4.2), determine the principal eigenvector (Section 4.1) or solve the optimization problem (Sections 4.3, 4.4), this might be a viable and handy solution.

Summing up, according to the proposed solution, one should: complete the missing elements according to the formula (4.13) and then compute the ranking as for the complete matrix using e.g., GMM (Section 3.2).

4.5.2 Ranking based on spanning trees

Every PC matrix C can be unambiguously represented in the form of a graph $P_C = (A, E)$ (Section 2.4) and let $S_C = (A, Q')$ be a spanning tree of G_C. It is easy to see that each such tree immediately determines a certain ranking of alternatives [126, 105, 193]. Indeed, it is enough to assign $w(a_1) = 1$ then adopt $w(a_p) = c_{1i_1}c_{i_1i_2} \cdot \ldots \cdot c_{i_{p-1}i_p}$ where $a_1, a_{i_1}, a_{i_2} \ldots, a_p$ forms the path between a_1 and a_p in P_C. For example, let us consider the following incomplete PC matrix

$$
M = \begin{pmatrix}
1 & 2 & 3 & ? \\
\frac{1}{2} & 1 & \frac{1}{4} & ? \\
\frac{1}{3} & 4 & 1 & \frac{1}{2} \\
? & ? & 2 & 1
\end{pmatrix}.
$$

There are three spanning trees of P_M: $S_1 = (A, \{\{a_1, a_2\}, \{a_1, a_3\}, \{a_3, a_4\}\})$, $S_2 = (A, \{\{a_1, a_2\}, \{a_2, a_3\}, \{a_3, a_4\}\})$ and $S_3 = (A, \{\{a_1, a_3\}, \{a_2, a_3\}, \{a_3, a_4\}\})$, where $A = \{a_1, \ldots, a_4\}$. The rankings determined by S_1, S_2 and S_3 (Chapter 2.4) are

$$
w_1 = \begin{pmatrix} 1 \\ 2 \\ 3 \\ \frac{3}{2} \end{pmatrix}, \quad w_2 = \begin{pmatrix} 1 \\ 2 \\ \frac{1}{2} \\ \frac{1}{4} \end{pmatrix}, \quad w_3 = \begin{pmatrix} 1 \\ 3 \\ 12 \\ 6 \end{pmatrix}.
$$

After rescaling so that the sum of their entries equals 1, we obtain

$$
\widehat{w}_1 = \begin{pmatrix} \frac{2}{15} \\ \frac{4}{15} \\ \frac{2}{5} \\ \frac{1}{5} \end{pmatrix}, \quad \widehat{w}_2 = \begin{pmatrix} \frac{4}{15} \\ \frac{8}{15} \\ \frac{2}{15} \\ \frac{1}{15} \end{pmatrix}, \quad \widehat{w}_3 = \begin{pmatrix} \frac{1}{22} \\ \frac{3}{22} \\ \frac{6}{11} \\ \frac{3}{11} \end{pmatrix}.
$$

Based on the individual spanning tree rankings, *Siraj et al.* [179, p. 194] proposed the priority vector, which is the arithmetic or geometric mean of all individual vectors, i.e.,

$$
w = \sum_{i=1}^{\eta} \widehat{w}_i,
$$

or

$$
w = \prod_{i=1}^{\eta} \widehat{w}_i,
$$

where η is the number of all possible spanning trees for the given PC matrix. When the matrix is complete, there are n^{n-2} such trees. Hence, the practical application of this method is limited to matrices not greater than 9×9 [179, p. 197]. The second method, GMAST – geometric mean of all spanning trees, however, turned out to be equivalent to the GMM method [126] (Section 3.2). This correspondence is also maintained for incomplete matrices [25]. This means that GMAST for incomplete PC matrices is equivalent to LLSM (Section 4.3) and GMM (Section 4.2) for incomplete PC matrices.

Chapter 5

Rating Scale

5.1 The concept of a rating scale

The rating scale is probably the most controversial and debatable element of the AHP method. According to *Saaty's* original proposition (see the fundamental scale, Section 1.2.4), experts assess alternatives using linguistic terms, such as weak, strong or important etc. For example, when comparing two alternatives, the expert does not answer the question whether the first is twice as important as the second, but whether the first is slightly more important than the second. Later on, comparative phrases containing words such as slightly, strong, moderate and so on, are automatically translated into numbers. For example, the phrase that a_i is slightly more important than a_j can result in the assignment $c_{ij} = 2$. Why should the word slightly correspond to a double difference between the alternatives? The theory does not answer that. It is easy to imagine that the waiting time for a meeting with a friend extended by one minute can be called in relation to the assumed one-minute time "slightly elongated." However, if your colleague earns twice as much as you do, you will not say that he earns "slightly more." Thus, it is evident that the actual meanings of the linguistic terms vary depending on the context. This is just one example of the problems with the concept of scale in AHP.

In fact, arbitrary and fixed assignment of linguistic terms to numbers is one of the most criticized elements of *Saaty's* method [55, 80, 84]. People understanding the same words differently, regularity and discreteness of the fundamental scale may lead to the creation of an artificial (i.e., caused by the scale itself) inconsistency.

Scale type	Form of transformation	Examples
Absolute scale	$f(x) = x$ (identity)	counting
Ratio scale	$f(x) = \alpha \cdot x, \ \alpha > 0$	mass, time, brightness (brils)
Interval scale	$f(x) = \alpha \cdot x + \beta, \ \alpha > 0$	temperature (centigrade), time
Ordinal scale	$f(x) \geq f(y) \ \textit{iff} \ x \succeq y$	preferences, hardness, air quality
Nominal scale	f – any bijection	labeling, indexing, etc.

Table 5.1: Types of scales

From a practical point of view, however, the use of a linguistic scale is quite convenient. Following *Barzilai* [15], let us recall that the scale f is a mapping of the objects in X into the objects in Y that reflects the structure of X into Y. To define f formally, let us define the notion of a relational system [152, 167, p. 51].

Definition 24. *Let $r_1, r_2, \ldots, r_p$ be relations in the same set X and $o_1, o_2, \ldots, o_q$ be binary operations in X. The relational system in X is defined as the $p + q + 1$ tuple in the form $\mathcal{X} = (X, r_1, \ldots, r_p, o_1, \ldots, o_q)$, where $p \geq 1$ and $q \geq 0$.*

$\mathcal{X}$ is called a numerical relational system if X is a set of real numbers. For example, $(\mathbb{R}, >, +)$ is a numerical relational system. The mapping that preserves the relational structure between relational systems is a homomorphism. Formally, homomorphisms can be defined as follows.

Definition 25. *Let $\mathcal{X}$ and $\mathcal{Y}$ be two relational systems $\mathcal{X} = (X, r_1, \ldots, r_p, o_1, \ldots, o_q)$ and $\mathcal{Y} = (Y, \widehat{r}_1, \ldots, \widehat{r}_p, \widehat{o}_1, \ldots, \widehat{o}_q)$. The mapping $f : X \to Y$ is said to be a homomorphism between $\mathcal{X}$ and $\mathcal{Y}$ if, for all $x_1, \ldots, x_k \in X$, the fact that $(x_1, \ldots, x_k) \in r_i$ implies that $(f(x_1), \ldots, f(x_k)) \in \widehat{r}_i$, and similarly for any two x_i and x_j from X it holds that $f(x_i o_k x_j) = f(x_i) \widehat{o}_k f(x_j)$.*

The scale is said to be a homomorphism between any relational system and the numerical relational system. *Roberts* [152, p. 64] indicates five basic types of scales (Table 5.1).

Despite the fact that we need to indicate the domain, co-domain and mapping between them to define the scale formally, the concept of scale is usually identified with the mapping itself. Thus, in Table 5.1, the "form of transformation" column indicates the functions characteristic for a given type of scale. For AHP, and generally for the pairwise comparisons method, two of the above scales are particularly important. The first one is the ratio scale. Assuming that the expert (or decision maker) is able to assess alternatives using the ratio scale, we can safely assign priorities to alternatives and compare one to the other. For example, if we ask someone how long bridge i is, then we ask how long bridge j is and the answer we get is in steps, then we clearly get two numbers in the form $\alpha \cdot$ |length of i-th bridge| and $\alpha \cdot$ |length of j-th bridge|. Hence, regardless of the α (which may correspond to the length of the step in meters of the person being asked for the opinion), if we compare these two values and we get a result, e.g., 2, it is clear that, according to the expert, the i-th bridge is two times longer than the j-th bridge. As you can see, the ratio scale is "convenient" in the context of pairwise comparisons. If the expert makes a comparison of alternatives with the ratio scale in mind, we can safely perform many operations on such numbers, including multiplication, division and scaling, etc.

The matter becomes complicated when experts start using the ordinal scale. In the original AHP, experts are asked to judge the alternatives in terms of the fundamental scale of absolute numbers (Table 1.1). Thus, considering the given comparison, an expert decides which of two alternatives is better, then assigns to the winner a verbal expression describing its relative importance. In this approach, the defeated alternative becomes the reference point "the unit of comparisons" [163], while the winning alternative is evaluated in terms of the fundamental scale. Thus, for example, an expert provides us with the judgment: $y =$ 'reference point', $x =$ 'moderately more important than the reference point', then, following the fundamental scale, we transform this information into the number $1/3$, i.e., we assign 1 to y and 3 to x. This type of transformation corresponds to the ordinal scale mapping (Table 5.1). The question then arises as to whether such a transformation makes sense. In fact, the question can be divided into three sub-questions. The first concerns the scale values $1, \ldots, 9$. Why are these the subsequent natural numbers from 1 to 9? The second concerns verbal expressions, such as weak, moderate or strong. Why should we use, for example, the term "strong" but not "severe"? Finally, even if we agree on some set of verbal expressions, like weak, moderate or strong, why should we map "weak importance" to 2, "moderate importance" to 3 and so on? Studying the works written by *Saaty*, one can only find the answer to the first of these questions. *Saaty* indicates that when considering comparisons of stimuli with intensities subject to Weber's law and noticeable differences, we get the sequence of integers $1, 2, 3, \ldots$ and so on [163, p .258]. Such an approach requires, in particular, the assumption that an expert's opinion can be quantified by the same rules as, for example, a feeling of pain. The answer to the question whether this is the case goes beyond mathematics, and therefore this study. The second and third questions are related to each other. Both concern the relative importance assigned to verbal expressions. This relativism means that various experts can make this assignment in various ways. This may lead, at least, to the incomparability of results.

Problems with the fundamental scale mean that experts are sometimes asked to express their judgments directly using real numbers. This eliminates the step of converting verbal expressions into numbers. It allows us to expect that the result of the comparison really means the actual ratio between the importance of the first and the second alternatives. Is it enough to cope with the concept of scale for AHP? *Barzilai* indicates that the scale as defined by *Roberts* [152] may be insufficient [15].

Despite formal doubts, the use of the fundamental scale and other verbal scales in combination with AHP are quite popular. In the next section, several other scales proposed as alternatives to the fundamental scale will be discussed.

5.2   Rating scales for AHP

5.2.1   Rating scale in AHP

The problem with the formal definition of the concept of scale in AHP has been addressed in [159]. In this approach, the scale is a mapping between a set of comparisons and real numbers. Hence, any transformation $P : A^2 \to \mathbb{R}_+$, where $A = \{a_1, \ldots, a_n\}$ – is a set of alternatives, such that if $a_i \succ a_j$ (a_i is more preferred than a_j) then $P(a_i, a_j) > 1$ and $a_i \sim a_j$ (a_i is indifferent to a_j) then $P(a_i, a_j) = 1$ is called a scale. In fact, Saaty calls every such mapping a fundamental (or primitive) scale [159], however, it was assumed to be called a fundamental scale only for such a mapping[1] with a co-domain consisting of integers $1, \ldots, 9$. It is easy to see that, in this approach, the scale is not a homomorphism as A^2 is not a relational system (we can consider the precedence relation with respect to alternatives but not to comparisons). Hence, the notion of a primitive scale as introduced by *Saaty* is not a concept of the primitive scale devised by Roberts [152]. Both of them, however, understand the concept of a derived scale as "a mapping between two numerical relational systems." Indeed, several derived scales built on top of *Saaty's* primitive scale will be shown in this chapter. *Saaty* does not raise the problem of mapping verbal terms to the numbers either. They are somehow hidden in the concept of P and do not appear in the definition of the scale P explicitly. The scale P is considered as a ratio scale [159, 77]. This means that the comparisons c_{ij} should be treated quantitatively with all the consequences of that fact. In this light, the mapping between verbal expressions and integers defined by a fundamental scale (Table 1.1) boils down to a convenient method of initial rounding (approximation) of real values related to alternatives in order to calculate their ratio.

---

[1]Very often, in practice, the term scale applies only to the co-domain of this mapping, i.e., the set of numbers, e.g., $1-5$. That is why the fundamental scale is very often defined as just a set of numbers $1, 2, \ldots, 9$.

5.2.2 Primitive scales

In addition to the fundamental scale [155], *Saaty* considers other primitive scales. They map a set of comparisons to subsequent positive integers $1 - 5$, $1 - 7$, $1 - 15$, $1 - 20$, and $1 - 90$. In the article, they are used to analyze inconsistency. In particular, it can be seen that the larger the scale range, the more inconsistent the random matrix. *Harker* and *Vargas* [77] consider primitive scales $1 - 5$ and $1 - 15$. They claim, using the example of distance measurements between cities [157, 74], that the fundamental scale performs better than its competitors.

5.2.3 Power scales

Harker and *Vargas* also considered two derived, power scales over the fundamental scale, given as the functions $f_2, f_{\sqrt{}} : \{1, \ldots, 9\} \to \mathbb{R}_+$. These are $f_2(x) = x^2$ and $f_{\sqrt{}}(x) = \sqrt{x}$. The values of the first one are $\{1, 4, 9, 16, 25, 36, 49, 64, 81\}$ and the second are $\{1, \sqrt{2}, \sqrt{3}, 2, \sqrt{5}, \sqrt{6}, \sqrt{7}, 2\sqrt{2}, 3\}$. Although *Harker* and *Vargas* [77] claimed that both of them perform worse than the fundamental scale, the Montecarlo study provided by *Dong* et al. [51] clearly indicates that f_2 results in the smallest RMS (root-mean-square deviation) out of all considered scales.

5.2.4 Balanced scale

The balanced scale proposal [173] comes from *Salo* and *Hämäläinen*. It is based on the observation made by *Schoner* and *Wedley* [175]. These researchers noticed that AHP produces consistently less accurate estimates, especially when the compared alternatives (the areas covered by colors) are similar. *Salo* and *Hämäläinen's* explained this phenomenon by the uneven dispersion of the local priorities when the scale is $r = \{1/9, \ldots, 1, \ldots, 9\}$ and the priorities of the two compared alternatives are given as $w(a_i) = 1/(r + 1)$ and $w(a_i) = r/(r + 1)$ (Fig. 5.1).

Therefore, they proposed the use of the inverse relationship $r = w/(1 - w)$ where w takes regularly distant values. For example, the values $0.05, 0.1, 0.15, \ldots, 0.9$ translate into the scale $\{0.052, 0.11, 0.176, 0.25, 0.33, 0.428, 0.538, 0.66, 0.818, 1, 1.22, 1.5, 1.857, 2.33, 3, 4, 5.66, 9\}$. This approach results in the regular distribution of the ratios (Fig. 5.2). It is worth noting that w can take different values. Because the fundamental scale understood as $1/9, 1/8, \ldots, 1/2, 1, 2, \ldots, 9$ is composed of 17 elements then as w we may adopt 17 points regularly distributed within the interval $[0.1, 0.9]$, i.e., $1/10, 3/20, \ldots, 9/10$. This results in the scale $\{0.11, 0.176, 0.25, 0.333\ 0.428, 0.538, 0.66, 0.818, 1, 1.22, 1.5, 1.857, 2.333, 3, 4, 5.66, 9\}$. Although the authors use the interval $[0.1, 0.9]$, they note that any interval strictly included in $(0, 1)$ might also be used instead of it.

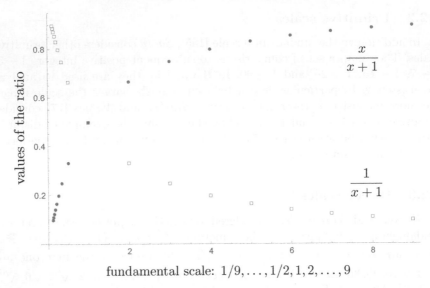

fundamental scale: $1/9, \ldots, 1/2, 1, 2, \ldots, 9$

Figure 5.1: Uneven dispersion of local priorities. The number of possible local priorities (results of comparisons) around the value of 0.5 is less than 0.1 or 0.9

According to *Dong* [51], the balanced scale performs better than *Saaty's* scale, but slightly worse than f_2.

5.2.5　Geometric scale

In [125], *Lootsma* proposes the geometric scale defined as the mapping $f_g(x) = \lambda^{x-1}$ where λ is a scale parameter and x is the appropriate integer. The inspiration to propose this scale was the observation, well-known in psychological measurement [152], that the difference between the subsequent grades e_k and e_{k-1} of a scale must be greater than or equal to the smallest noticeable difference, which is proportional to the smaller of the two. This leads to the assumption that the following relationship $e_k = \epsilon \cdot e_{k-1}$ takes place between the grades of a scale. Thus, in particular $e_k = (1 + \epsilon)^k e_0$. By adopting $e_0 = 1$ and $\lambda = (1 + \epsilon)$ we obtain $f_g(x)$. *Franek* and *Kresta* [65] use the geometric scale with $\epsilon = 1$. *Finan* and *Hurely* [62] consider several variants of *Lootsma's* scale with $\epsilon = 0.2, 0.4, \ldots, 2$.

5.2.6　Other scales

There are several other scales that have not been mentioned before. There include the inverse linear scale

$$f_{\mathrm{mz}}(x) = \frac{9}{10 - x}, \quad \text{for } x = 1, \ldots, 9$$

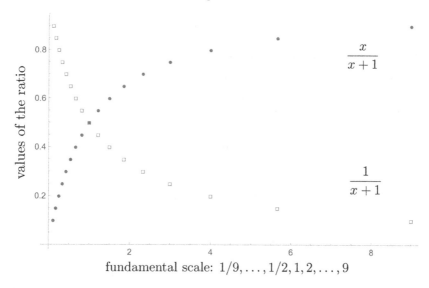

Figure 5.2: Regular dispersion of local priorities for the balanced scale

proposed by *Ma* and *Zheng* [127]. This scale also performed better in an experiment conducted by *Dong* et al. [51]. Another scale is *Dodd* and *Donegan's* asymptotical scale [50, 88, 65]:

$$f_{dd}(x) = \tanh^{-1}\left(\frac{\sqrt{3}(x-1)}{14}\right) \quad \text{for } x = 1,\ldots,9.$$

The third is the logarithmic scale devised by *Ishizaka* et al. [87]. According to this approach, the subsequent grades of a scale take the values of the binary logarithm of natural numbers $1,\ldots,9$. Formally, the logarithmic scale can be written as

$$f_{ibk}(x) = \log_2(x+1), \quad \text{for } x = 1,\ldots,9.$$

The above three transformations lead to the following degrees of scale,

$$f_{mz} : 1, 1.125, 1.28571, 1.5, 1.8, 2.25, 3, 4.5, 9$$

$$f_{dd} : 0, 0.124, 0.253, 0.39, 0.542, 0.723, 0.955, 1.32, 2.634$$

$$f_{ibk} : 1, 1.58, 2, 2.322, 2.585, 2.807, 3, 3.170, 3.322$$

All these scales are more dense around 1 and the further they are from one, the less frequently their grades are located.

Koczkodaj [99] argues that the relative priority is insufficient to determine the actual mutual priority of the two alternatives. He indicates that, in addition to the ratio of alternatives, the scale upper limit must also be known. One solution is to recreate the scale so that all its grades fit the interval $[0, 1]$.

5.3   Which scale to choose

Using scale in the context of AHP is controversial[2]. Issues such as the selection of linguistic expressions, transformation between linguistic expressions and numbers, formal definition of the scale, formal problems (including the lack of zero on the scale [50]) and distribution of numerical degrees of the scale cause vigorous debate among researchers.

Some completely reject AHP as useless; *"The fundamental mathematical error of using inapplicable operations to construct AHP scales renders the numbers generated by the AHP meaningless"* [15], others accept it as it is. Some researchers use directly numerical continuous ratio scales. This, of course, requires more effort connected with experts explaining the interpretation of the numbers they provide, and the efforts of the experts themselves resulting from the evaluation of alternatives. Thanks to this, some of the scale problems seem to disappear. However, it is no longer AHP as such, but a variation of it. There is also a question as to whether it is not a simple transfer of the problem to the expert (or decision-maker).

However, if we want or need to use the scale of verbal expressions, the question arises as to which scale we should choose. In general, different scales for AHP were proposed due to various imperfections of the fundamental scale. These imperfections could be observed in a certain context, hence the choice of some scale should be well suited to the situation for which it was defined. An exception may be the power scale f_2. According to *Dong* et al. [51], this scale performs better than the competitors under random circumstances. It is therefore a viable alternative to the fundamental scale.

---

[2]The rating scale problem was the inspiration underlying the creation of MACBETH method [12, 61, p. 410].

Chapter 6

Inconsistency

6.1   Introduction

As we could observe in the chapter on prioritization methods for incomplete PC matrices (Chapter 4), the $n \times n$ PC matrix contains more results of pairwise comparisons than necessary to calculate the ranking. Indeed, the minimal number of comparisons is $n-1$ providing, of course, that the matrix is irreducible. However, more than the minimum number of comparisons can be used to verify the quality of the data collected. What does this verification consist of? In principle, we may expect that every single entry c_{ij} of a PC matrix C corresponds to the priority ratio of the i-th and j-th alternatives, i.e.,

$$c_{ij} = \frac{w(a_i)}{w(a_j)}. \tag{6.1}$$

In particular, one can expect that

$$c_{ik} = \frac{w(a_i)}{w(a_k)} \text{ and } c_{kj} = \frac{w(a_k)}{w(a_j)},$$

which leads to the conclusion that

$$c_{ij} = c_{ik} \cdot c_{kj} \tag{6.2}$$

for every $i, j, k = 1, \ldots, n$. In other words, the value of the ratio $w(a_i)/w(a_j)$ can be determined directly as c_{ij} and indirectly as $c_{ik} \cdot c_{kj}$. Moreover, it is easy to show that any product in the form $c_{ik_1} c_{k_1 k_2} \cdot \ldots \cdot c_{k_{r-1} k_r} c_{k_r j}$ also equals c_{ij}. Since every single value of c_{ij} is the result of experts' judgment, it is possible that, due to imprecise assessment or scale discretization, the individual value c_{ij} only approximates the actual ratio $w(a_i)/w(a_j)$. Therefore it is possible that there are such i, k and j that

$$c_{ij} \neq c_{ik} \cdot c_{kj}. \tag{6.3}$$

Of course, there may be even longer products of comparisons, such that

$$c_{ij} \neq c_{ik_1} c_{k_1 k_2} \cdot \ldots \cdot c_{k_{r-1} k_r} c_{k_r j}.$$

The discrepancies between c_{ij} and products that we expect should be equal to c_{ij} are the basis for identifying and determining the inconsistency of a PC matrix. Methods for measuring the degree of inconsistency are called inconsistency indices.

However, the question arises as to the importance of the inconsistency. Is it good if a PC matrix is fairly consistent, or is it bad if it is inconsistent? In AHP or, more generally, in the pairwise comparisons method, it is assumed that the experts make decisions rationally. Therefore, they will try to make the results of pair comparisons as close as possible to the appropriate ratios of priorities.

Therefore, we may expect that the results of comparison provided by a fair and reliable expert will be quite consistent. High inconsistency may therefore mean that either the expert did not behave rationally (he/she was driven by emotions or mood), did not exercise due diligence in making the assessment, or simply did not know the subject of judgment. In each of these cases, it is doubtful whether the ranking calculated on this basis is trustworthy. Therefore, in the method of pairwise comparisons, it is usual to assume that high inconsistency means that the data is unreliable and it is not worth using them to calculate the ranking.

However, one may ask whether a high or even fully consistent PC matrix gives a guarantee that the ranking is correct. The brief answer is no. *Forman,* in his article [64] on facts and fictions about AHP, explicitly states, *"It is important that a low inconsistency does not become the goal of the decision-making process. A low inconsistency is a necessary but not a sufficient condition for a good decision. It is more important to be accurate than consistent. It is possible to be perfectly consistent but consistently wrong."* In other words, it is better to have a fairly consistent set of data than a highly inconsistent one, but in extreme cases experts might be consistently wrong. Consistency is not guaranteed by the expertise of experts. Conversely, it is also possible that rankings based on inconsistent data will be completely correct.

6.2 Preliminaries

One of the basic ways of calculating and detecting inconsistencies is triad analysis. Let us formally define the notion of a triad.

Definition 26. *A group of three distinct alternatives $\{a_i, a_j, a_k\}$ is called a triad if $\{i, k, j\} \in 2^{\{1,...,n\}}$, where n is the total number of alternatives. Very often, it is convenient to call a triad the set of indices $\{i, j, k\}$ corresponding to these alternatives.*

As mentioned before, inconsistency is a situation in which the value of comparison between two alternatives determined in several ways is different. In the simplest case, it is enough to consider three comparisons to detect inconsistency (6.3). Therefore, the simplest definition of inconsistency is based on the notion of a triad – the results of three pairwise comparisons.

Definition 27. *An $n \times n$ PC matrix $C = [c_{ij}]$ is said to be inconsistent if there is a triad $\{i, j, k\}$ such that $c_{ij} \neq c_{ik}c_{kj}$. Otherwise C is said to be consistent.*

The assumption made at the beginning of this chapter that when every single comparison c_{ij} equals the ratio $w(a_i)/w(a_j)$ then the matrix is consistent works both ways. This observation leads to the following theorem.

Theorem 4. *Let $w : A \to \mathbb{R}_+$ be a priority function assigning priorities to alternatives and $C = [c_{ij}]$ be an $n \times n$ PC matrix. In that case, wherever C is consistent then for every $i, j = 1, \ldots, n$ it holds that $c_{ij} = w(a_i)/w(a_j)$.*

Proof. "$\Leftarrow$" Let us assume that for every $i, j = 1, \ldots, n$ it holds that

$$c_{ij} = \frac{w(a_i)}{w(a_j)}. \tag{6.4}$$

Then, in particular, it is true that for every $k \in \{1, \ldots, n\}$ and $k \neq i, j$

$$c_{ik} = \frac{w(a_i)}{w(a_k)} \quad \text{and} \quad c_{kj} = \frac{w(a_k)}{w(a_j)},$$

hence

$$w(a_i) = c_{ik} w(a_k) \quad \text{and} \quad w(a_k) = c_{kj} w(a_j).$$

Thus, substituting one into another

$$w(a_i) = c_{ik} c_{kj} w(a_j).$$

Hence, due to (6.4) we obtain

$$c_{ij} = \frac{c_{ik} c_{kj} w(a_j)}{w(a_j)},$$

which means

$$c_{ij} = c_{ik} c_{kj}$$

for every $i, j = 1, \ldots, n$. $\qquad\square$

Proof. "$\Rightarrow$" Let us assume that for every $i, j = 1, \ldots, n$ it holds that $c_{ij} = w(a_i)/w(a_j)$. For the purpose of the proof by contradiction, let us assume that C is inconsistent, i.e., there exists a triad $\{i, j, k\}$ such that $c_{ij} \neq c_{ik} c_{kj}$. Hence,

$$c_{ij} \neq \frac{w(a_i)}{w(a_k)} \cdot \frac{w(a_k)}{w(a_j)} = \frac{w(a_i)}{w(a_j)}.$$

Thus,

$$w(a_i) \neq c_{ij} w(a_j).$$

Contradiction. $\qquad\square \qquad\qquad\qquad\qquad\qquad \square$

It is worth noting that the above theorem does not depend on the choice of the mapping w.

Property 1. *This leads to the conclusion that, for a consistent PC matrix, the rankings calculated using different priority-deriving methods must be identical with respect to the constant scaling factor.*

In other words, if $C = [c_{ij}]$ is consistent, and

$$w_x = \begin{pmatrix} w_x(a_1) \\ \vdots \\ \vdots \\ w_x(a_n) \end{pmatrix}, \quad w_y = \begin{pmatrix} w_y(a_1) \\ \vdots \\ \vdots \\ w_y(a_n) \end{pmatrix}$$

are the two priority vectors obtained by the two different priority methods X and Y, then there exists $\alpha \in \mathbb{R}$ such that $w_x = \alpha w_y$. Indeed, since for a consistent PC matrix it holds that:

$$\frac{w_x(a_i)}{w_x(a_j)} = c_{ij} = \frac{w_y(a_i)}{w_y(a_j)}$$

then

$$w_x(a_i) = \frac{w_x(a_j)}{w_y(a_j)} \cdot w_y(a_i).$$

As the above equation does not depend on the selection of i, then it must also hold for every $j = 1, \ldots, n$. In other words

$$\frac{w_x(a_1)}{w_y(a_1)} = \ldots = \frac{w_x(a_j)}{w_y(a_j)} = \ldots = \frac{w_x(a_n)}{w_y(a_n)}.$$

Hence, assuming $\alpha = w_x(a_j)/w_y(a_j)$ we obtain the desired property $w_x = \alpha w_y$.

Property 2. *Consistency also implies that one column of a PC matrix can be obtained from another column by multiplication with a scalar.*

Indeed, let

$$q_i = \begin{pmatrix} c_{1i} \\ c_{2i} \\ \vdots \\ c_{ni} \end{pmatrix}, \quad q_j = \begin{pmatrix} c_{1j} \\ c_{2j} \\ \vdots \\ c_{nj} \end{pmatrix},$$

be the i-th and j-th columns of $C = [c_{ij}]$. Then, directly from (Def. 27) it holds that

$$c_{ij} = c_{ik} c_{kj}$$

for $k = 1, \ldots, n$, and, of course:

$$\frac{1}{c_{ik}} = \frac{1}{c_{ij}} c_{kj}.$$

However, due to reciprocity, the above equation can be written as

$$c_{ki} = c_{ji} c_{kj}.$$

Due to the free choice of k, $q_i = c_{ji}q_j$, i.e., the i-th column can be obtained from the j-th column by multiplying the latter by c_{ji}. This, of course, implies that the determinant of a consistent PC matrix is 0.

There is also another immediate observation connected with the columns of a consistent PC matrix.

Property 3. *Every column of a consistent PC matrix equals the ranking with respect to the constant scaling factor.*

This can be simply verified and the i-th column can be written as

$$q_i = \begin{pmatrix} c_{1i} \\ c_{2i} \\ \vdots \\ c_{ni} \end{pmatrix} = \begin{pmatrix} \frac{w(a_1)}{w(a_i)} \\ \frac{w(a_2)}{w(a_i)} \\ \vdots \\ \frac{w(a_n)}{w(a_i)} \end{pmatrix}.$$

Thus, $\alpha \cdot q_i$ equals the priority vector w, where $\alpha = w(a_i)$. In other words, for a consistent PC matrix there is no need to calculate the priority vector. It is enough to take any column and multiply it by the chosen real and positive constant.

6.3 Quantifying inconsistency

In practice, a fully consistent matrix is quite a rare case. Data from experts are not always accurate. Nevertheless, we must be able to assess how far the experts were inaccurate in their judgments. In AHP, methods for assessing this inaccuracy are called inconsistency (or consistency) indices. It seems that the inconsistency indices can be divided into two major groups: the ranking-dependent indices and the matrix-based indices [119].

The first group is composed of indices which are based on the assumption that the ratio $w(a_i)/w(a_j)$ should be similar and sometimes even equal to the corresponding entry c_{ij}. Thus, taking into account the difference between $w(a_i)/w(a_j)$ and c_{ij} and the number of violations of equality $w(a_i)/w(a_j) = c_{ij}$, we should be able to determine the extent to which the given $C = [c_{ij}]$ is inconsistent. A characteristic feature of this group of indices is that, for their calculation, it is necessary to calculate the ranking of alternatives beforehand. In this chapter, we will deal with the following indices belonging to this group:

- Saaty's consistency index,

- Geometric consistency index,

- Golden–Wang index,

- Relative error,

- Maximal error,

- Logarithmic least square criterion.

The second group is composed of all the indices which do not need the ranking to be computed to determine inconsistency. Most of them are based on the notion of a triad (Def. 26). Thus, they directly need the PC matrix and do not care about the ranking as such. Of course, this division is conventional in the sense that the ranking is calculated based on the PC matrix. Hence, ultimately, each index is calculated based on a pairwise comparison matrix. In this chapter, we will deal with the following indices belonging to this group:

- Koczkodaj inconsistency index,

- Index of determinants,

- Triad-based average inconsistency indices,

- Ambiguity index,

- Cavallo and D'Apuzzo index.

The above-mentioned indices do not exhaust the list. At the end of this section, reference will be made to other inconsistency indices.

6.3.1 Saaty's consistency index

The most popular and well-known inconsistency measure is Saaty's consistency index CI. This is an example of an index based on ranking (here EVM). It is defined as:

$$CI(C) \overset{df}{=} \frac{\lambda_{max} - n}{n - 1}, \tag{6.5}$$

where λ_{max} is a principal eigenvalue of the PC matrix C.

Example 7. *To determine $CI(C)$ one needs to compute the principal eigenvalue. This is not easy and it is best to use a calculation package or spreadsheet. For example, when*

$$C = \begin{pmatrix} 1 & 7 & \frac{1}{6} & \frac{1}{2} & \frac{1}{4} & \frac{1}{6} & 4 \\ \frac{1}{7} & 1 & \frac{1}{3} & 5 & \frac{1}{5} & \frac{1}{7} & 5 \\ 6 & 3 & 1 & 6 & 3 & 2 & 8 \\ 2 & \frac{1}{5} & \frac{1}{6} & 1 & 8 & \frac{1}{5} & 8 \\ 4 & 5 & \frac{1}{3} & \frac{1}{8} & 1 & \frac{1}{9} & 2 \\ 6 & 7 & \frac{1}{2} & 5 & 9 & 1 & 2 \\ \frac{1}{4} & \frac{1}{5} & \frac{1}{8} & \frac{1}{8} & \frac{1}{2} & \frac{1}{2} & 1 \end{pmatrix} \tag{6.6}$$

then the principal eigenvalue of C is $\lambda_{max} = 11.2344$ and $CI(C) = 0.705$.

6.3.1.1 Meaning of the index

By this formula, *Saaty* tried to capture both the intensity of inconsistency, measured as the difference between $w(a_i)/w(a_j)$ and c_{ij}, and the number of cases in which $w(a_i)/w(a_j) \neq c_{ij}$. To see this [155, p. 238], let us look at the eigenvalue equation in the form:

$$Cw = \lambda_{max} w.$$

In particular, it means that

$$\sum_{j=1}^{n} c_{ij} w(a_j) = \lambda_{max} w(a_i),$$

i.e.,

$$\sum_{j=1}^{n} c_{ij} \frac{w(a_j)}{w(a_i)} = \lambda_{max}.$$

As $c_{ii} = w(a_i)/w(a_i) = 1$ then

$$\sum_{\substack{j=1 \\ i \neq j}}^{n} c_{ij} \frac{w(a_j)}{w(a_i)} = \lambda_{max} - 1.$$

Since the above is true for every $i = 1, \ldots, n$ then

$$\sum_{i=1}^{n} \sum_{\substack{j=1 \\ i \neq j}}^{n} c_{ij} \frac{w(a_j)}{w(a_i)} = n\lambda_{max} - n,$$

which might be written as

$$\frac{1}{n} \sum_{1 \leq i < j \leq n}^{n} \left(c_{ij} \frac{w(a_j)}{w(a_i)} + c_{ji} \frac{w(a_i)}{w(a_j)} \right) = \lambda_{max} - 1.$$

Thus, by subtracting $n - 1$ from both sides of the equation and dividing by $n - 1$ we obtain

$$-1 + \frac{1}{n(n-1)} \sum_{1 \leq i < j \leq n}^{n} \left(c_{ij} \frac{w(a_j)}{w(a_i)} + c_{ji} \frac{w(a_i)}{w(a_j)} \right) = \frac{\lambda_{max} - n}{n - 1}. \tag{6.7}$$

The value $c_{ij} w(a_i)/w(a_j)$ may be treated as a kind of error indicator or a discrepancy measure between the matrix and the ranking. Let us denote

$$e_{ij} \stackrel{df}{=} c_{ij} \frac{w(a_j)}{w(a_i)}. \tag{6.8}$$

Thus, (6.7) can be written as

$$CI(C) = -1 + \frac{1}{n(n-1)} \sum_{1 \leq i < j \leq n}^{n} (e_{ij} + e_{ji}). \tag{6.9}$$

Since the function $x + 1/x$ has a minimum for $x = 1$, then also $e_{ij} + e_{ji} = e_{ij} + e_{ij}^{-1}$ is minimal when $e_{ij} = 1$, i.e., when $c_{ij} = w(a_i)/w(a_j)$. Hence, $CI(C)$ is minimal when for every $i, j = 1, \ldots, n$ it holds that $c_{ij} = w(a_i)/w(a_j)$, thus due to Theorem 4, C is consistent. In such a case $e_{ij} + e_{ij}^{-1} = 2$, i.e.,

$$CI(C) = -1 + \frac{1}{n(n-1)} \sum_{1 \leq i < j \leq n}^{n} 2,$$

i.e.,

$$CI(C) = -1 + \frac{1}{n(n-1)} \frac{n(n-1)}{2} 2 = 0.$$

Equation (6.9) clearly shows that $CI(C)$ is greater with larger individual expressions $e_{ij} + e_{ij}^{-1}$ and more such expressions that $e_{ij} + e_{ij}^{-1} > 2$. In other words, Saaty's consistency index takes into account the number and size of individual consistency violation considered as $e_{ij} \neq 1$, for $i, j = 1, \ldots, n$. So, despite the rather enigmatic formula (6.5), its mathematical meaning is quite clear and simple.

6.3.1.2 Reciprocity need

Formula (6.9) also implies that $CI(C) \geq 0$. However, it is easy to see that if C is not reciprocal, then $CI(C)$ may get a value smaller than 0. For example, if

$$M = \begin{pmatrix} 1 & 2 & 3 & 4 \\ \frac{1}{2} & 1 & 1 & 2 \\ \frac{1}{3} & \frac{2}{3} & 1 & \frac{4}{3} \\ \frac{1}{4} & \frac{1}{2} & \frac{3}{4} & 1 \end{pmatrix},$$

then $CI(M) = -0.028 < 0$. Because a negative value of the inconsistency index is meaningless, CI is not suitable for determining the inconsistency of the non-reciprocal PC matrices. Indeed, AHP assumes that the PC matrix must be reciprocal. This assumption has the form of an axiom [159, p. 844].

6.3.1.3 Improving consistency

The error, as defined above (6.8), allows users to identify and gradually correct the inconsistency [169, p. 8,9]. The consistency improvement procedure relies on identifying the highest error value e_{ij} and asking the expert (or experts) to change the value of the problematic comparison c_{ij} so that it would be closer to the corresponding ratio $w(a_i)/w(a_j)$. If the expert agrees to change his mind, the new value c_{ij} translates into a lower value of $CI(C)$.

6.3.1.4   Upper limit of inconsistency

Besides the inappropriate use of *CI* for non-reciprocal PC matrices, its value is equal or greater than 0. However, the question arises as to how large $CI(C)$ can be, and as follows, what the most inconsistent PC matrix is. *Aupetit* and *Genest*[1] [7] showed that for a primitive scale of $1, 2, \ldots, q$ the most inconsistent PC matrix takes the form

$$
C_q = \begin{pmatrix}
1 & q & \frac{1}{q} & q & \cdots \\
\frac{1}{q} & 1 & q & \frac{1}{q} & \cdots \\
q & \frac{1}{q} & 1 & q & \cdots \\
\frac{1}{q} & q & \frac{1}{q} & 1 & \cdots \\
\vdots & \vdots & \vdots & \vdots & \ddots
\end{pmatrix}.
$$

Since $C_q = [c_{ij}]$ is a positive matrix, the principal eigenvalue λ_{max} of C_q is the *Perron–Frobenius* eigenvalue. Thus, it meets the following inequality [129]:

$$
\min_j \sum_{i=1}^{n} c_{ij} \leq \lambda_{max} \leq \max_j \sum_{i=1}^{n} c_{ij}. \tag{6.10}
$$

It is easy to see that if $n > 1$ is odd, i.e., $n = 2r + 1$ where $r \in \mathbb{N}_+$, then

$$
\min_j \sum_{i=1}^{n} c_{ij} = \max_j \sum_{i=1}^{n} c_{ij},
$$

and

$$
\max_j \sum_{i=1}^{n} c_{ij} = 1 + \underbrace{q + \ldots + q}_{(n-1)/2} + \underbrace{\frac{1}{q} + \ldots + \frac{1}{q}}_{(n-1)/2} = 1 + \frac{n-1}{2}\left(q + \frac{1}{q}\right).
$$

Thus, due to (6.10), it holds that:

$$
\lambda_{max} = 1 + \frac{n-1}{2}\left(q + \frac{1}{q}\right).
$$

For the same reason, the consistency index $CI(C_q)$ where n is odd is given as:

$$
CI(C_q) = \frac{\lambda_{max} - n}{n - 1} = \frac{(q-1)^2}{2q} \tag{6.11}
$$

Therefore, assuming $q = 9$ as in the fundamental scale, Saaty's consistency index equals $CI(C_q) \approx 3.55556$, where C_q is an $n \times n$ PC matrix and n is odd.

---

[1]Čerňanová et al. [38] called this matrix (LPM) the layer-cake PC matrix.

Now, we will show that for any positive and reciprocal $C_q = [c_{ij}]$, where $c_{ij} \in [\frac{1}{q}, q]$, the consistency index CI can never be greater than $CI(C_q)$ and, as follows, C_q is the most inconsistent judgment matrix with respect to CI whose entries belong to $\{\frac{1}{q}, \frac{1}{q-1}, \ldots, \frac{1}{2}, 1, 2, \ldots, q-1, q\}$. In particular, it would mean that in AHP with the fundamental scale Saaty's consistency index cannot be larger than $32/9 \approx 3.55556$.

To prove the desired property, let us consider the auxiliary Lemma 1, then Theorem 5.

Lemma 1. *The sum of principal eigenvalues of the matrices A and A^T is not greater than the principal eigenvalue of $\widehat{A} = A + A^T$.*

Proof. Let λ_A, λ_{A^T} and $\lambda_{\widehat{A}}$ denote the principal eigenvalues of A, A^T and $\widehat{A} = A + A^T$, correspondingly. Let v_A be a principal eigenvector of A scaled so that its length is 1, i.e., $\|v_A\| = 1$. In such a case $v_A^T v_A = 1$. Thus, due to the *Rayleigh quotient*[2] [146, p. 12],

$$\lambda_A = v_A^T M v_A = \langle M v_A, v_A \rangle$$

and similarly,

$$\lambda_{A^T} = v_{A^T}^T M^T v_{A^T} = \langle M^T v_{A^T}, v_{A^T} \rangle.$$

Hence

$$\lambda_A + \lambda_{A^T} = (\langle M v_A, v_A \rangle + \langle M^T v_A, v_A \rangle) = \langle (M + M^T) v_A, v_A \rangle.$$

Therefore

$$\lambda_A + \lambda_{A^T} \leq \sup_{\|v\|=1} \langle (M + M^T) v, v \rangle = \lambda_{\widehat{A}}$$

which completes the proof[3]. $\square$

Theorem 5. *The value of Saaty's consistency index $CI(C)$ for the pairwise comparisons matrix $C = [c_{ij}]$ built over the scale $1, \ldots, q$ is not greater than $CI(C_q)$, i.e.,*

$$CI(C) \leq \frac{(q-1)^2}{2q}.$$

Proof. Let us consider the PC matrix $C = [c_{ij}]$ in the form:

$$C = \begin{pmatrix} 1 & c_{12} & \cdots & c_{1n} \\ \frac{1}{c_{12}} & 1 & \cdots & c_{2n} \\ \vdots & \cdots & \ddots & \vdots \\ \frac{1}{c_{1n}} & \cdots & \cdots & 1 \end{pmatrix}.$$

---

[2] $\langle u, v \rangle$ denotes a scalar product (dot product) of the vectors u and v.
[3] Thanks to professor *Ryszard Szwarc* for help in the proof.

such that $c_{ij} \in [\frac{1}{q}, q]$ and $q > 0$. Let λ_C denote the principal eigenvalue of C. By analogy, its transposition is given as

$$
C^T = \begin{pmatrix}
1 & \frac{1}{c_{12}} & \cdots & \frac{1}{c_{1n}} \\
c_{12} & 1 & \cdots & \frac{1}{c_{2n}} \\
\vdots & & \ddots & \vdots \\
c_{1n} & \cdots & \cdots & 1
\end{pmatrix}
$$

and let λ_{C^T} denote the principal eigenvalue of C^T. Both C and C^T have the same sets of eigenvalues, thus in particular it holds that $\lambda_C = \lambda_{C^T}$. The sum of C and C^T is given as:

$$
\widehat{C} = C + C^T = \begin{pmatrix}
2 & c_{12} + \frac{1}{c_{12}} & \cdots & c_{1n} + \frac{1}{c_{1n}} \\
c_{12} + \frac{1}{c_{12}} & 2 & \cdots & c_{2n} + \frac{1}{c_{2n}} \\
\vdots & & \ddots & \vdots \\
c_{1n} + \frac{1}{c_{1n}} & \cdots & \cdots & 2
\end{pmatrix}
$$

and let $\lambda_{\widehat{C}}$ denote the principal eigenvalue of $\widehat{C}$. Similarly as in (6.10) $\lambda_{\widehat{C}}$ is bounded by the minimal and maximal sum of the rows, i.e.,

$$
2 + \min_j \sum_{\substack{i=1, \\ i \neq j}}^n \left(c_{ij} + \frac{1}{c_{ij}} \right) \leq \lambda_{\widehat{C}} \leq 2 + \max_j \sum_{\substack{i=1, \\ i \neq j}}^n \left(c_{ij} + \frac{1}{c_{ij}} \right). \tag{6.12}
$$

It easy to notice that the function $f(x) = x + \frac{1}{x}$ for $x \in [\frac{1}{q}, q]$, where $q \in \mathbb{R}_+$ has two equivalent maxima in $x = \frac{1}{q}$ and $x = q$, and one minimum in $x = 1$. Hence,

$$
2 + (n-1)f(1) \leq \lambda_{\widehat{C}} \leq 2 + (n-1)f(q),
$$

i.e.,

$$
2 + 2(n-1) \leq \lambda_{\widehat{C}} \leq 2 + (n-1)\left(q + \frac{1}{q} \right). \tag{6.13}
$$

Due to Lemma 1 it holds that $\lambda_C + \lambda_{C^T} \leq \lambda_{C+C^T}$. Since $\lambda_C = \lambda_{C^T}$, equation (6.13) can be written as:

$$
2n \leq 2\lambda_C \leq \lambda_{\widehat{C}} \leq 2 + (n-1)\left(q + \frac{1}{q} \right).
$$

In particular

$$
n \leq \lambda_C \leq 1 + \frac{1}{2}(n-1)\left(q + \frac{1}{q} \right).
$$

By subtracting n and dividing by $n-1$ we obtain

$$
0 \leq CI(C) \leq \frac{(q-1)^2}{2q}
$$

which completes the proof. $\qquad\qquad\qquad\qquad\qquad\qquad\qquad\qquad\qquad\square$

One of the conclusions of the above theorem is the observation that the end of the inconsistencies of a given matrix does not depend on the size of the matrix, but on the scale range. This leads to an immediate observation that, in order to ensure an inconsistency smaller than the given value $CI(C) \leq x$, it would be enough for the maximum comparison c to satisfy inequality

$$x \leq \frac{(c-1)^2}{2c},$$

i.e.,

$$0 \leq c^2 - 2c(1+x) + 1.$$

This means that if for every comparison c_{ij} it holds that

$$\frac{1}{\sqrt{x^2 + 2x} + x + 1} \leq c_{ij} \leq \sqrt{x^2 + 2x} + x + 1,$$

then $CI(C) \leq x$. For example, every PC matrix $C = [c_{ij}]$ such that $0.641742 \leq c_{ij} \leq 1.55826$ for $i, j = 1, \ldots, n$ has an inconsistency smaller than 0.1. The above example shows that the average inconsistency is relative and depends on the scale. Therefore, when assessing the value of inconsistency, we must always remember about the adopted range of judgment scale. An attempt to make the concept of inconsistency independent of the adopted scale is the consistency ratio considered in the next section 6.3.2.

6.3.2 Consistency ratio

The value of the consistency index depends on the size of the PC matrix. The same error in a small matrix will result in greater inconsistency than in a large matrix. This results from equation (6.9) on the basis of which it is easy to see that CI is an average of individual errors, given as $e_{ij} + e_{ji}$. Hence, the more components included in the mean, the less important a single element is. CI is also sensitive to the scale range. It turns out that the larger the scale range, the more individual errors appear to increase the inconsistency. Considering this, Saaty suggested computing the average CI value for random matrices of different sizes and scales [155] and using it to scale the consistency index. Let us denote this value as $RI_{n,q}$, where n is the size of a matrix and q is the range of a scale. When carrying out the Monte Carlo experiment, it is easy to see that indeed the scale and size of a matrix affects inconsistency (Fig. 6.1, Table 6.1).

$RI_{n,q}$ is used as the reference point for the consistency index CI. Hence, in order to minimize the effect of the matrix size and the scale range on CI, Saaty suggests using the consistency ratio:

$$CR(C) \overset{df}{=} \frac{CI(C)}{RI_{n,q}}, \tag{6.14}$$

where C is an $n \times n$ PC matrix over the primitive scale $1, \ldots, q$. The value $CR(C)$ indicates how much more inconsistent C is in comparison to a random

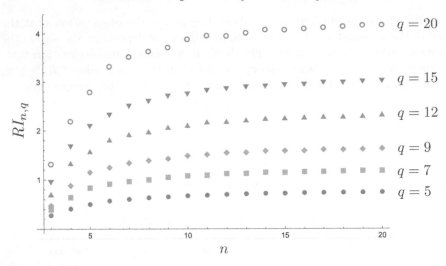

Figure 6.1: Average value of CI for PC matrices of different sizes $n \times n$ and primitive scales $1, \ldots, q$

matrix with the same size and scale. For example, if $CR(C) = 0.5$ it means that C is on average two times more consistent than the corresponding random matrix. Based on $CR(C)$, Saaty proposed to treat all the PC matrices with the inconsistency CI at least ten times smaller than the random case as sufficiently consistent. As a result, each matrix for which $CR(C) \leq 0.1$ was considered reliable, and the other matrices not meeting this condition were considered to be in need of improvement. Despite the high popularity of 0.1 as the acceptability threshold for the $CR(C)$, we must remember that this choice is arbitrary. Therefore, depending on the specific situation and circumstances, it is worth considering other $CR(C)$ acceptability thresholds. For example, the value of the consistency ratio for the PC matrix used in the example (7) is

$$CR(C) = \frac{CI(C)}{RI_{7,9}} = \frac{0.705}{1.33} = 0.53.$$

6.3.3   Geometric consistency index

The geometric consistency index (GCI) has been proposed by *Aguaròn* and *Moreno-Jimènez* [3]. The idea is based on the original proposal from *Crawford* and *Williams* [46]. Similarly to Saaty's consistency index, GCI also uses the error e_{ij} (6.8). It is easy to see that for the consistent PC matrix C every $e_{ij} = 1$ and, as follows, $\log e_{ij} = 0$. This observation gave *Aguaròn* and *Moreno-Jimènez* the idea to use an unbiased estimator of the variance of perturbations [3] as the inconsistency measure for C. To avoid negative values, they use a squared logarithm, i.e., $\log^2 e_{ij}$ instead of $\log e_{ij}$ (the expression

	n=3	n=4	n=5	n=6	n=7	n=8	n=9	n=10	n=11	n=12	n=13	n=14	n=15	n=16	n=17	n=18	n=19	n=20
q=3	0.109	0.186	0.223	0.250	0.27	0.285	0.294	0.298	0.306	0.31	0.318	0.320	0.323	0.327	0.327	0.33	0.333	0.334
q=4	0.181	0.3	0.367	0.405	0.434	0.448	0.477	0.485	0.49	0.5	0.507	0.509	0.52	0.524	0.524	0.528	0.53	0.534
q=5	0.245	0.406	0.506	0.552	0.602	0.633	0.657	0.667	0.681	0.698	0.707	0.716	0.719	0.731	0.73	0.736	0.739	0.744
q=6	0.306	0.515	0.64	0.722	0.782	0.825	0.846	0.872	0.887	0.895	0.917	0.924	0.935	0.936	0.946	0.952	0.955	0.962
q=7	0.376	0.632	0.786	0.905	0.972	1.02	1.04	1.07	1.1	1.11	1.12	1.14	1.15	1.15	1.17	1.17	1.17	1.18
q=8	0.495	0.762	0.933	1.09	1.15	1.20	1.24	1.28	1.3	1.33	1.34	1.35	1.36	1.38	1.38	1.39	1.4	1.4
q=9	0.546	0.83	1.08	1.26	1.33	1.41	1.45	1.47	1.51	1.54	1.55	1.57	1.58	1.59	1.61	1.61	1.62	1.63
q=10	0.570	1.03	1.27	1.41	1.5	1.6	1.66	1.7	1.75	1.76	1.77	1.79	1.8	1.82	1.83	1.84	1.85	1.85
q=11	0.639	1.18	1.42	1.6	1.74	1.79	1.87	1.9	1.94	1.97	1.99	2.015	2.019	2.041	2.058	2.069	2.072	2.087
q=12	0.831	1.23	1.58	1.78	1.93	2	2.084	2.137	2.166	2.178	2.211	2.244	2.256	2.267	2.294	2.297	2.299	2.311
q=13	0.776	1.37	1.78	1.95	2.099	2.197	2.278	2.33	2.361	2.407	2.423	2.462	2.472	2.496	2.511	2.523	2.534	2.547
q=14	0.860	1.49	1.9	2.175	2.3	2.405	2.503	2.539	2.591	2.627	2.67	2.681	2.71	2.726	2.734	2.748	2.767	2.781
q=15	0.915	1.58	2.041	2.296	2.491	2.615	2.675	2.759	2.826	2.844	2.891	2.909	2.933	2.94	2.963	2.981	3	3.006
q=16	0.986	1.7	2.258	2.515	2.709	2.805	2.93	2.983	2.996	3.075	3.098	3.119	3.156	3.174	3.189	3.205	3.223	3.238
q=17	1.07	1.81	2.399	2.637	2.88	3.023	3.122	3.194	3.238	3.274	3.322	3.36	3.383	3.4	3.452	3.437	3.457	3.461
q=18	1.15	2.064	2.575	2.817	3.109	3.235	3.306	3.394	3.474	3.481	3.562	3.582	3.609	3.637	3.655	3.668	3.696	3.704
q=19	1.19	2.132	2.705	3.101	3.251	3.425	3.552	3.63	3.668	3.719	3.767	3.793	3.852	3.858	3.876	3.899	3.913	3.935
q=20	1.34	2.28	2.876	3.197	3.48	3.634	3.732	3.824	3.907	3.959	3.999	4.023	4.059	4.091	4.109	4.134	4.158	4.176

Table 6.1: $RI_{n,q}$ values. Each value is an average consistency index of 1000 random matrices of size $n \times n$ over the primitive scale $1, 2, \ldots, q$.

$\log^2 e_{ij}$ can be treated as the distance $(\log e_{ij} - \log 1)^2$. Thus, the devised formula takes the form:

$$I_G(C) \stackrel{df}{=} \frac{2}{(n-1)(n-2)} \sum_{i=1}^{n} \sum_{j=i+1}^{n} \log^2 e_{ij}.$$

In fact, I_G is the rescaled mean of the expressions $\log^2 e_{ij}$ and can be written as

$$I_G(C) = \frac{n}{n-2} \cdot \frac{1}{\binom{n}{2}} \sum_{i=1}^{n} \sum_{j=i+1}^{n} \log^2 e_{ij}.$$

In other words, GCI is the rescaled mean of logarithmized and squared errors of all the comparisons above the diagonal of C where the scaling factor depends on the size of the matrix and is given as $n/(n-2)$.

Example 8. *Let us calculate I_G for*

$$C = \begin{pmatrix} 1 & 2 & \frac{1}{2} & \frac{1}{9} \\ \frac{1}{2} & 1 & 5 & 9 \\ 2 & \frac{1}{5} & 1 & \frac{1}{6} \\ 9 & \frac{1}{9} & 6 & 1 \end{pmatrix}.$$

The geometric (non-rescaled) rank is given as $w = \left[\frac{1}{\sqrt{3}}, \sqrt[4]{\frac{5}{2}}\sqrt{3}, \frac{1}{\sqrt[4]{15}}, \sqrt[4]{6} \right]^T$.
Thus, the geometric consistency index is given as

$$I_G(C) = \frac{1}{3} \left(\log^2 \left(2 \cdot 3 \sqrt[4]{\frac{5}{2}} \right) + \log^2 \left(\frac{1}{2} \cdot \sqrt[4]{\frac{3}{5}} \right) + \log^2 \left(\frac{1}{9} \cdot \sqrt[4]{2} \cdot 3^{3/4} \right) + \right.$$

$$\left. + \log^2 \left(5 \cdot \frac{\sqrt[4]{2}}{3^{3/4}\sqrt{5}} \right) + \log^2 \left(9 \cdot \frac{\sqrt{2}}{\sqrt[4]{15}} \right) + \log^2 \left(\frac{1}{6} \cdot \sqrt{3}\sqrt[4]{10} \right) \right).$$

After calculating the above formula, we get $I_G(C) = 3.38359$.

6.3.4 Golden–Wang index

The *Golden–Wang* inconsistency index [73] is based on the observation that, for a consistent PC matrix (Property 3), every column creates the ranking itself. In other words, the quantitative relations between alternatives determined by a column are exactly the same as in the final priority vector. Thus, if for a consistent PC matrix, one would rescale both the priority vector and its single column, so that all their entries sum up to one, then both are identical. This property can also be used for inconsistent matrices. It turns out that the more inconsistent the PC matrix, the greater the difference between one and the other.

Let $C^* = [c_{ij}^*]$ be a PC matrix obtained from $C = [c_{ij}]$ by scaling the columns of C so that every column sums up to 1, i.e.,

$$c_{ij}^* = \frac{c_{ij}}{\sum_{k=1}^{n} c_{kj}}.$$

Similarly, let w_{gmm} be a rescaled GMM priority vector (3.13). Then, the Golden–Wang consistency index is defined as the average distance between the rescaled columns and the priority vector,

$$GW(C) \overset{df}{=} \frac{1}{n} \sum_{i=1}^{n} \sum_{j=1}^{n} \mid c_{ij}^* - w_{gmm}(a_i) \mid. \tag{6.15}$$

It is worth knowing that, in the original work [73], GMM is proposed as a way of calculating the priority vector used in the above formula (6.15). In the light of Theorem 4, however, any priority-deriving method based on the assumption that $w(a_i)/w(a_j)$ tends toward c_{ij} can be used.

Example 9. *Let us consider the PC matrix*

$$C = \begin{pmatrix} 1 & 4 & \frac{1}{3} & \frac{1}{4} \\ \frac{1}{4} & 1 & \frac{1}{4} & 1 \\ 3 & 4 & 1 & \frac{1}{4} \\ 4 & 1 & 4 & 1 \end{pmatrix}.$$

It is easy to calculate that

$$C^* = \begin{pmatrix} \frac{4}{33} & \frac{2}{5} & \frac{4}{67} & \frac{1}{10} \\ \frac{1}{33} & \frac{1}{10} & \frac{3}{67} & \frac{2}{5} \\ \frac{4}{11} & \frac{2}{5} & \frac{12}{67} & \frac{1}{10} \\ \frac{16}{33} & \frac{1}{10} & \frac{48}{67} & \frac{2}{5} \end{pmatrix}, \quad w_{gm} = \begin{pmatrix} 0.166 \\ 0.109 \\ 0.287 \\ 0.437 \end{pmatrix},$$

where w_{gm} is the priority vector obtained by using the geometric mean method. Thus, the Golden–Wang inconsistency index for C is given as follows:

$$GW(C) = \frac{1}{4} \left(\left| \frac{4}{33} - 0.166 \right| + \left| \frac{2}{5} - 0.166 \right| + \ldots + \left| \frac{2}{5} - 0.437 \right| \right) = 0.52.$$

6.3.5 Relative error

Barzilai's proposal [14, 119] is devised for additive PC matrices. Let $\widehat{C} = [\widehat{c}_{ij}]$ be an n by n additive PC matrix. and let Δ_i denote the sum of its i-th row, i.e., $\Delta_i = \frac{1}{n}\sum_{j=1}^{n}\hat{c}_{ij}$. Hence, $X = [x_{ij}]$ will be the auxiliary matrix where x_{ij} is the difference between the sums of the i-th and j-th columns, i.e.,

$$x_{ij} = \Delta_i - \Delta_j. \tag{6.16}$$

The second auxiliary matrix $E = [\epsilon_{ij}]$ is defined as

$$\epsilon_{ij} = \hat{c}_{ij} - x_{ij}. \tag{6.17}$$

The relative error is given as follows:

$$RE(\hat{C}) \stackrel{df}{=} \frac{\sum_{ij}\epsilon_{ij}^2}{\sum_{ij}\hat{c}_{ij}^2}. \tag{6.18}$$

The relative error measure is accompanied by the relative consistency index defined as:

$$RC(\hat{C}) \stackrel{df}{=} \frac{\sum_{ij}x_{ij}^2}{\sum_{ij}\hat{c}_{ij}^2}.$$

Both concepts complement each other so that they sum up to 1, i.e.,

$$RE(\hat{C}) + RC(\hat{C}) = 1.$$

Although the above notions were defined for additive PC matrices, they can also be used for multiplicative PC matrices. To do this, one needs to use logarithmic transformation. Thus, for the multiplicative PC matrix $C = [c_{ij}]$ and its additive counterpart $\hat{C} = [\log c_{ij}]$ we obtain:

$$RE(C) \stackrel{df}{=} RE(\hat{C}), \text{ and } RC(C) \stackrel{df}{=} RC(\hat{C}).$$

6.3.6   Logarithmic least square criterion

In the chapter dedicated to priority-deriving methods, optimization methods (Sec. 3.3) occupy an important place. The idea behind them is to minimize a certain expression $\mathcal{E}$ involving ratios in the form $w(a_i)/w(a_j)$ and the results of comparisons given as c_{ij}. The priority values $w(a_i)$ for $i, j = 1, \ldots, n$ play the role of variables. The result of the optimization process are optimal values of $w(a_i)$ where $i, j = 1, \ldots, n$. These optimal values minimize $\mathcal{E}$. The expression being optimized, however, can be considered as the inconsistency index. Indeed, it is easy to prove that for a consistent PC matrix the minimal value for $\mathcal{E}$ is 0. The most popular representative of optimization methods is the logarithmic least square method (LLSM). The minimization criterion in LLSM can be used as an inconsistency index [45, 23]. Let us recall it here.

$$\mathcal{E}(C) \stackrel{df}{=} \sum_{i=1}^{n}\sum_{j=1}^{n}\left(c_{ij} - \frac{w(a_i)}{w(a_j)}\right)^2.$$

Of course, the priority vector w used in the above equation is calculated either using LLSM or GMM.

6.3.7 Maximum error

Errors as defined in (6.8) are the elements that make up *Saaty's CI*. Following [110], let us look at the relationship between the errors e_{ij} and $K(C)$. To this end, we need to extend the error definition.

Definition 28. *Let the reciprocal error $\mathscr{E}$ be defined as:*

$$\mathscr{E}_{ij} \overset{df}{=} \max\{e_{ij} - 1, \frac{1}{e_{ij}} - 1\}. \tag{6.19}$$

It is easy to observe that if $e_{ij} \neq 1$ then $\mathscr{E}_{ij} > 0$. Let us define the maximal local inconsistency indicator.

Definition 29. *Let the maximal reciprocal error $\mathscr{E}_{max}(C)$ be defined as*

$$\mathscr{E}_{max}(C) \overset{df}{=} \max_{i,j} \mathscr{E}_{ij}.$$

$\mathscr{E}_{max}(C)$ is, in fact, just another inconsistency index. There is an interesting relationship between $\mathscr{E}_{max}(C)$ and $CI(C)$. Both indices have a common upper limit [109].

Theorem 6. *For the PC matrix C and the EVM priority vector w it holds that:*

$$\mathscr{E}_{max}(C) \leq \delta \Rightarrow CI(C) \leq \delta. \tag{6.20}$$

Proof. Since $\mathscr{E}_{max}(C) \leq \delta$, thus by definition it is true that $e_{ij} - 1 \leq \delta$ for every $i, j = 1, \ldots, n$. In particular, it holds that:

$$\sum_{\substack{i=1 \\ i \neq j}}^{n} (e_{ij} - 1) \leq (n-1)\delta.$$

Hence

$$\frac{1}{(n-1)} \sum_{\substack{i=1 \\ i \neq j}}^{n} (e_{ij} - 1) \leq \delta. \tag{6.21}$$

Since, according to *Saaty* [155]

$$CI(C) = \frac{1}{(n-1)} \sum_{\substack{i=1 \\ i \neq j}}^{n} (e_{ij} - 1), \tag{6.22}$$

then (6.21) means that the assertion $CI(C) \leq \delta$ is satisfied. $\qquad \square$

The above theorem shows that $\mathscr{E}_{max}$ can effectively replace *CI*. An additional value brought by $\mathscr{E}_{max}$ is the even distribution of local errors.

$\mathscr{E}_{max}(C)$ can be used to define the ranking effectiveness criterion [110], according to which the ranking procedure should respond to decreasing inconsistency with a decreasing error $\mathscr{E}_{max}(C)$.

Definition 30. *It is said that the prioritization method P is effective with respect to the inconsistency index $\mathcal{I}$ if for any reciprocal and multiplicative PC matrix C it holds that:*

$$\mathcal{I}(C) \to 0 \Rightarrow \mathscr{E}_{max}(C) \to 0$$

and if there is κ, where $\mathcal{I}(C) \geq \kappa \geq 0$, such that for every C':

$$\mathcal{I}(C) - \mathcal{I}(C') > \kappa \Rightarrow \mathscr{E}_{max}(C) - \mathscr{E}_{max}(C') > 0.$$

The effectiveness property formulates the natural expectations that, whenever we significantly reduce the inconsistency, the ranking error should also decrease. Later, the above definition will allow us to show that EVM is effective with respect to the Koczkodaj inconsistency index.

6.3.8   Koczkodaj inconsistency index

Koczkodaj's index [98, 53] is one of the simplest but, at the same time, one of the most useful indices of PC matrix inconsistency. On the one hand, it allows the inconsistency of the PC matrix to be determined while, on the other, it indicates the place which is the source of the highest inconsistency. It is convenient to divide its definition into two parts. The first concerns local inconsistency (triad inconsistency). Let the inconsistency for a triad a_i, a_k and a_j (triad inconsistency) be defined as

$$K_{i,j,k}(C) \stackrel{df}{=} \min\left\{\left|1 - \frac{c_{ik}c_{kj}}{c_{ij}}\right|, \left|1 - \frac{c_{ij}}{c_{ik}c_{kj}}\right|\right\}. \tag{6.23}$$

The global index is designed as the maximal local inconsistency. Thus, the index takes the form:

$$K(C) \stackrel{df}{=} \max\left\{K_{i,j,k}(C) \mid 1 \leq i < j < k \leq n\right\}. \tag{6.24}$$

The triad a_i, a_k, a_j for which $K_{i,j,k}(C) = K(C)$ is the most inconsistent triad in C. This information may be helpful in the process of improving the consistency of C in the same way as the local error e_{ij} (see Sec. 6.3.1.3). It is easy to prove that for every PC matrix C it holds that $0 \leq K(C) < 1$.

Example 10. *Let us consider the matrix*

$$C = \begin{pmatrix} 1 & \frac{69}{74} & \frac{10}{31} & \frac{7}{22} & \frac{19}{7} \\ \frac{74}{69} & 1 & \frac{27}{43} & \frac{2}{3} & \frac{34}{13} \\ \frac{31}{10} & \frac{43}{27} & 1 & \frac{47}{38} & \frac{73}{7} \\ \frac{22}{7} & \frac{3}{2} & \frac{38}{47} & 1 & 7 \\ \frac{7}{19} & \frac{13}{34} & \frac{7}{73} & \frac{1}{7} & 1 \end{pmatrix}.$$

The five alternatives considered introduce ten distinct triads, hence the set of subsets of indices is as follows

$$T = \{\{1,2,3\},\{1,2,4\},\{1,2,5\},\{1,3,4\},\{1,3,5\},$$

$$\{1,4,5\},\{2,3,4\},\{2,3,5\},\{2,4,5\},\{3,4,5\}\}.$$

Hence

$$K_{1,2,3}(C) = \frac{25933}{57753}, K_{1,2,4}(C) = \frac{247}{506}, K_{1,2,5}(C) = \frac{928}{9139}, K_{1,3,4}(C) = \frac{1047}{5170},$$

$$K_{1,3,5}(C) = \frac{141}{730}, K_{1,4,5}(C) = \frac{75}{418}, K_{2,3,4}(C) = \frac{539}{3807}, K_{2,3,5}(C) = \frac{15389}{25623},$$

$$K_{2,4,5}(C) = \frac{40}{91}, K_{3,4,5}(C) = \frac{471}{2774}.$$

The maximal triad inconsistency is $K_{2,3,5}(C)$. Thus, the Koczkodaj incon-sistency index for C is

$$K(C) = K_{2,3,5}(C) = \frac{15389}{25623} = 0.60059 \ .$$

6.3.8.1 Effectiveness of the Koczkodaj index

EVM and GMM are effective with respect to the *Koczkodaj's* index, i.e., the lower $K(C)$ the smaller $\mathscr{E}_{max}(C)$. Before we show that fact, let us prove a few interesting lemmas. The first one, however, has a technical meaning and it is needed in the proof of the second Lemma.

Lemma 2. *For every $x, y > 0$ holds that*

$$\alpha y \leq x \leq \frac{1}{\alpha}y,$$

where $\alpha = 1 - k$ and

$$k = \min\left\{\left|1 - \frac{x}{y}\right|, \left|1 - \frac{y}{x}\right|\right\}.$$

Proof. First, let us observe that

$$x < y \text{ implies that } k = 1 - \frac{x}{y}, \text{ and}$$

$$y < x \text{ implies that } k = 1 - \frac{y}{x}.$$

In order to prove the thesis let us assume for a moment that it does not hold. The first case we have to consider is the following:

$$x < \alpha y. \tag{6.25}$$

The above equation implies that

$$x < (1 - k)y. \tag{6.26}$$

Since, $\alpha < 1$ then due to (6.25) $x < y$. This means that the equation (6.26) can be written as

$$x < \left(1 - \left(1 - \frac{x}{y}\right)\right) y,$$

which means that

$$x < \left(\frac{x}{y}\right) y,$$

i.e.,

$$x < x$$

Contradiction. The thesis may not hold also when

$$\frac{1}{\alpha}y < x,$$

i.e.,

$$y < \alpha x.$$

Since $\alpha < 1$ then also $y < x$ the above equation can be written as

$$y < (1 - k)x = \left(1 - \left(1 - \frac{y}{x}\right)\right) x.$$

Thus,

$$y < \left(\frac{y}{x}\right) x$$

what once again leads to contradiction:

$$y < y.$$

□

Lemma 3. *For every reciprocal, multiplicative and complete PC matrix C and the priority vector w obtained by EVM, it is true that:*

$$\alpha \leq e_{ij} \leq \frac{1}{\alpha}, \tag{6.27}$$

where $\alpha \overset{df}{=} 1 - K(C)$ and $e_{ij} = c_{ij}w(a_j)/w(a_i)$.

Proof. According to (6.24), $K(C)$ is the maximal local $K_{i,j,k}(C)$, i.e., for any i, j, k it holds that $K(C) \geq K_{i,j,k}(C)$. In particular

$$K(C) \geq \min\left\{\left|1 - \frac{c_{ij}}{c_{ik}c_{kj}}\right|, \left|1 - \frac{c_{ik}c_{kj}}{c_{ij}}\right|\right\}.$$

Hence, either

$$c_{ij} \leq c_{ik}c_{kj} \text{ implies } K(C) \geq 1 - \frac{c_{ij}}{c_{ik}c_{kj}}$$

or

$$c_{ik}c_{kj} \leq c_{ij} \text{ implies } K(C) \geq 1 - \frac{c_{ik}c_{kj}}{c_{ij}}$$

is true. Let us denote $\alpha \overset{df}{=} 1 - K(C)$. The above expressions obtain the forms:

$$c_{ij} \leq c_{ik}c_{kj} \text{ implies } c_{ij} \geq \alpha \cdot c_{ik}c_{kj},$$

$$c_{ik}c_{kj} \leq c_{ij} \text{ implies } \frac{1}{\alpha} \cdot c_{ik}c_{kj} \geq c_{ij}$$

Since $\alpha \leq 1$, both of the above cases lead to the common observation (Lemma 2) that:

$$\alpha \cdot c_{ik}c_{kj} \leq c_{ij} \leq \frac{1}{\alpha}c_{ik}c_{kj} \text{ for } i,j = 1,\ldots,n. \tag{6.28}$$

As w is the (rescaled) principal eigenvector of C, then it satisfies the equation:

$$Cw = \lambda_{max}w,$$

where λ_{max} is the principal eigenvalue of C. It translates to the equation system where every single equation looks as follows

$$c_{j1}w(a_1) + \ldots + c_{jn}w(a_n) = \lambda_{max} \cdot w(a_j) \text{ for } j = 1,\ldots,n. \tag{6.29}$$

The error e_{ij} can be written as:

$$e_{ij} \overset{df}{=} \frac{1}{c_{ji}} \cdot \frac{w(a_j)}{w(a_i)} = \frac{1}{c_{ji}} \cdot \frac{c_{j1}w(a_1) + \ldots + c_{jn}w(a_n)}{c_{i1}w(a_1) + \ldots + c_{in}w(a_n)}. \tag{6.30}$$

Using (6.28) with (6.29), we obtain

$$c_{j1}w(a_1) + \ldots + c_{jn}w(a_n) \leq \frac{1}{\alpha}\left(c_{ji}c_{i1}w(a_1) + \ldots + c_{ji}c_{in}w(a_n)\right) \tag{6.31}$$

and

$$\alpha\left(c_{ji}c_{i1}w(a_1) + \ldots + c_{ji}c_{in}w(a_n)\right) \leq c_{j1}w(a_1) + \ldots + c_{jn}w(a_n). \tag{6.32}$$

Thus (6.31) means that

$$\frac{1}{c_{ji}} \cdot \frac{w(a_j)}{w(a_i)} \leq \frac{1}{c_{ji}} \cdot \left(\frac{1}{\alpha} \cdot \frac{c_{ji}\left(c_{i1}w(a_1) + \ldots + c_{in}w(a_n)\right)}{c_{i1}w(a_1) + \ldots + c_{in}w(a_n)}\right) = \frac{1}{\alpha},$$

and (6.32) implies that

$$\alpha = \frac{1}{c_{ji}} \cdot \left(\alpha \cdot \frac{c_{ji}\left(c_{i1}w(a_1) + \ldots + c_{in}w(a_n)\right)}{c_{i1}w(a_1) + \ldots + c_{in}w(a_n)} \right) \leq \frac{1}{c_{ji}} \cdot \frac{w(a_j)}{w(a_i)}.$$

Both of the above expressions lead to the observation that:

$$\alpha \leq e_{ij} \leq \frac{1}{\alpha}$$

which is the desired assertion. $\qquad\square$

Since $e_{ij} = e_{ji}^{-1}$ then by raising all the elements of (6.27) to -1 we get another important observation:

$$\alpha \leq e_{ji} \leq \frac{1}{\alpha}. \tag{6.33}$$

The above Lemma has an analogous version for the GMM.

Lemma 4. *For every reciprocal, multiplicative and complete PC matrix C and the priority vector w obtained by GMM, it is true that:*

$$\alpha \leq e_{ij} \leq \frac{1}{\alpha}, \tag{6.34}$$

where $\alpha \overset{df}{=} 1 - K(C)$ and $e_{ij} = c_{ij}w(a_j)/w(a_i)$.

Proof. Similarly as before (6.24) implies that $K(C)$ is the maximal local $K_{i,j,k}(C)$, i.e., for any i, j, k it holds that $K(C) \geq K_{i,j,k}(C)$. In particular for any triad in the form of c_{ik}, c_{kj}, c_{ij} it must hold that:

$$K(C) \geq \min\left\{ \left| 1 - \frac{c_{ij}}{c_{ik}c_{kj}} \right|, \left| 1 - \frac{c_{ik}c_{kj}}{c_{ij}} \right| \right\} \tag{6.35}$$

which, like in the previous reasoning, leads to the conclusion that:

$$\alpha \cdot c_{ik}c_{kj} \leq c_{ij} \leq \frac{1}{\alpha}c_{ik}c_{kj} \tag{6.36}$$

for every $i, j, k \in \{1, \ldots, n\}$.

On the other hand, wherever we consider ratio $w_{GM}(a_i)/w_{GM}(a_j)$ we may write it in the following form[4] [116]:

$$\frac{w(a_j)}{w(a_i)} = \frac{\left(\prod_{r=1}^n c_{jr}\right)^{\frac{1}{n}}}{\left(\prod_{r=1}^n c_{ir}\right)^{\frac{1}{n}}} = \frac{\left(\prod_{r=1}^n c_{jr}\right)^{\frac{1}{n}} \left(\prod_{r=1}^n w(a_r)\right)^{\frac{1}{n}}}{\left(\prod_{r=1}^n c_{ir}\right)^{\frac{1}{n}} \left(\prod_{r=1}^n w(a_r)\right)^{\frac{1}{n}}} = \frac{\left(\prod_{r=1}^n c_{jr}w(a_r)\right)^{\frac{1}{n}}}{\left(\prod_{r=1}^n c_{ir}w(a_r)\right)^{\frac{1}{n}}}.$$

---

[4] For simplicity, until the end of the proof we use w instead of w_{GM}.

Thus, for the purpose of comparing priorities we may assume that

$$w(a_i) = \left(\prod_{r=1}^{n} c_{ir} w(a_r) \right)^{\frac{1}{n}},$$

and still we get the same (up to some constant scaling factor) ranking as in the GMM approach. In other words e_{ij} can be written as:

$$e_{ij} \overset{df}{=} \frac{1}{c_{ji}} \cdot \frac{w(a_j)}{w(a_i)} = \frac{1}{c_{ji}} \cdot \frac{\left(\prod_{r=1}^{n} c_{jr} w(a_r) \right)^{\frac{1}{n}}}{\left(\prod_{r=1}^{n} c_{ir} w(a_r) \right)^{\frac{1}{n}}}. \tag{6.37}$$

Since, due to (6.36)

$$\left(\prod_{r=1}^{n} c_{jr} w(a_r) \right)^{\frac{1}{n}} \leq \left(\prod_{r=1}^{n} \frac{1}{\alpha} c_{ji} c_{ir} w(a_r) \right)^{\frac{1}{n}},$$

the left side of (6.37) leads us to the inequality:

$$\frac{1}{c_{ji}} \cdot \frac{\left(\prod_{r=1}^{n} c_{jr} w(a_r) \right)^{\frac{1}{n}}}{\left(\prod_{r=1}^{n} c_{ir} w(a_r) \right)^{\frac{1}{n}}} \leq \frac{1}{c_{ji}} \cdot \frac{\left(\prod_{r=1}^{n} \frac{1}{\alpha} c_{ji} c_{ir} w(a_r) \right)^{\frac{1}{n}}}{\left(\prod_{r=1}^{n} c_{ir} w(a_r) \right)^{\frac{1}{n}}}. \tag{6.38}$$

Since,

$$\frac{1}{c_{ji}} \cdot \frac{\left(\prod_{r=1}^{n} \frac{1}{\alpha} c_{ji} \right)^{\frac{1}{n}} \left(\prod_{r=1}^{n} c_{ir} w(a_r) \right)^{\frac{1}{n}}}{\left(\prod_{r=1}^{n} c_{ir} w(a_r) \right)^{\frac{1}{n}}} = \frac{1}{c_{ji}} \cdot \left(\frac{1}{\alpha} c_{ji} \right) = \frac{1}{\alpha}, \tag{6.39}$$

then

$$e_{ij} \leq \frac{1}{\alpha}. \tag{6.40}$$

Similarly, due to (6.36)

$$\left(\prod_{r=1}^{n} \alpha c_{ji} c_{ir} w(a_r) \right)^{\frac{1}{n}} \leq \left(\prod_{r=1}^{n} c_{jr} w(a_r) \right)^{\frac{1}{n}},$$

we may rewrite the left side of (6.37) in the form of inequality:

$$\frac{1}{c_{ji}} \cdot \frac{\left(\prod_{r=1}^{n} \alpha c_{ji} c_{ir} w(a_r) \right)^{\frac{1}{n}}}{\left(\prod_{r=1}^{n} c_{ir} w(a_r) \right)^{\frac{1}{n}}} \leq \frac{1}{c_{ji}} \cdot \frac{\left(\prod_{r=1}^{n} c_{jr} w(a_r) \right)^{\frac{1}{n}}}{\left(\prod_{r=1}^{n} c_{ir} w(a_r) \right)^{\frac{1}{n}}}. \tag{6.41}$$

Since,

$$\frac{1}{c_{ji}} \cdot \frac{\left(\prod_{r=1}^{n} \alpha c_{ji} \right)^{\frac{1}{n}} \left(\prod_{r=1}^{n} c_{ir} w(a_r) \right)^{\frac{1}{n}}}{\left(\prod_{r=1}^{n} c_{ir} w(a_r) \right)^{\frac{1}{n}}} = \frac{1}{c_{ji}} \cdot (\alpha c_{ji}) = \alpha, \tag{6.42}$$

then

$$\alpha \leq e_{ij}. \tag{6.43}$$

Both the above inequalities (6.40) and (6.43) lead to the conclusion that:

$$\alpha \leq e_{ij} \leq \frac{1}{\alpha}, \tag{6.44}$$

where the priority vector was computed using the geometric mean method. $\square$

Lemmas 3 and 4 together with (6.33) will be used to prove the effectiveness property. Following [110], let us prove the following theorem.

Theorem 7. *EVM and GMM are effective with respect to Koczkodaj inconsistency index $K(C)$.*

Proof. Based on Lemmas 3 and 4

$$\alpha - 1 \leq e_{ij} - 1 \leq \frac{1}{\alpha} - 1 \text{ for } i, j = 1, \ldots, n,$$

where e_{ij} is based on either EVM- or GMM-based ranking vector (6.30, 6.37), and due to (6.33) it holds that

$$\alpha - 1 \leq \frac{1}{e_{ij}} - 1 \leq \frac{1}{\alpha} - 1.$$

Thus,

$$0 \leq \max\{e_{ij} - 1, \frac{1}{e_{ij}} - 1\} \leq \frac{1}{\alpha} - 1. \tag{6.45}$$

In other words (see 6.19):

$$0 \leq \mathscr{E}_{ij} \leq \frac{1}{\alpha} - 1 \text{ for } i, j = 1, \ldots, n. \tag{6.46}$$

Then, due to (Def. 29), it holds that:

$$0 \leq \mathscr{E}_{max}(C) \leq \frac{1}{\alpha} - 1. \tag{6.47}$$

As $\alpha = 1 - K(C)$, then

$$K(C) \to 0 \text{ implies } \alpha \to 1 \text{ and } \left(\frac{1}{\alpha} - 1\right) \to 0. \tag{6.48}$$

Therefore, due to (6.47)

$$K(C) \to 0 \quad \text{implies} \quad \mathscr{E}_{max}(C) \to 0 \tag{6.49}$$

which satisfies the first claim of the effectiveness property (Def. 30).

To prove the second claim of the effectiveness property, let us look back at its right side once again.

Let us assume for a while that

$$\mathscr{E}_{max}(C') \le \left(\frac{1}{1 - K(C')} - 1 \right) < \mathscr{E}_{max}(C). \tag{6.50}$$

It is true if and only if

$$K(C') < 1 - \frac{1}{\mathscr{E}_{max}(C) + 1}.$$

The above is equivalent to

$$K(C) - K(C') > K(C) + \frac{1}{\mathscr{E}_{max}(C) + 1} - 1. \tag{6.51}$$

Thus, let κ be defined as

$$\kappa \overset{df}{=} K(C) + \frac{1}{\mathscr{E}_{max}(C) + 1} - 1. \tag{6.52}$$

In other words, the second claim of the effectiveness property is true if only for $0 \le \kappa \le K(C)$.

The fact that for an inconsistent PC matrix $\kappa \le K(C)$ results from the observation that $\mathscr{E}_{max}(C) \ge 0$, i.e., the expression

$$\frac{1}{\mathscr{E}_{max}(C) + 1} - 1 \le 0.$$

On the other hand, (6.47) implies that

$$\mathscr{E}_{max}(C) \le \frac{1}{1 - K(C)} - 1$$

thus,

$$1 - \frac{1}{\mathscr{E}_{max}(C) + 1} \le K(C)$$

and finally

$$0 \le K(C) + \frac{1}{\mathscr{E}_{max}(C) + 1} - 1,$$

i.e.,

$$0 \le \kappa.$$

Thus, for the κ as defined above (6.52) it holds that $0 \le \kappa \le K(C)$ and

$$K(C) - K(C') > \kappa \Leftrightarrow \mathscr{E}_{max}(C) - \mathscr{E}_{max}(C') > 0.$$

$\square$

The above theorem provides a guarantee that, whenever we reduce the value of the Koczkodaj inconsistency index, we also reduce the maximal error $\mathscr{E}_{ij}$ providing, of course, that the priority vector is calculated using EVM.

In addition to effectiveness, Lemma 3 provides estimation for the ranking values. Hence, the immediate observation is that, through $K(C)$, the priority of the i-th alternative is bounded by the priority of the j-th alternative.

Corollary 1. *In EVM and GMM, for priorities of any two alternatives it holds that*

$$\alpha c_{ij} w(a_j) \le w(a_i) \le \frac{1}{\alpha} c_{ij} w(a_j) \tag{6.53}$$

where $\alpha = 1 - K(C)$.

In particular, the above observation shows that the lower the inconsistency the closer $w(a_i)$ and $c_{ij} w(a_j)$. In particular, if $K(C) = 0$ then $w(a_i) = c_{ij} w(a_j)$ for any $i, j = 1, \ldots, n$.

6.3.8.2　Koczkodaj index and CI

In a similar way to how $K(C)$ was used to limit the maximal error in the PC matrix under EVM, the Koczkodaj index can also be used to determine the limit of the consistency index $CI(C)$ [110].

Theorem 8. *For any PC matrix C it holds that*

$$CI(C) \le \frac{1}{1 - K(C)} - 1 \tag{6.54}$$

Proof. Following *Saaty* [155], the consistency index can also be written as:

$$\frac{1}{(n-1)} \sum_{\substack{i=1 \\ i \ne j}}^{n} (e_{ij} - 1) = CI(C) \tag{6.55}$$

Based on Lemma 3, we know that

$$\alpha \le e_{ij} \le \frac{1}{\alpha}$$

thus,

$$\alpha - 1 \le e_{ij} - 1 \le \frac{1}{\alpha} - 1.$$

Thus,

$$(n-1)(\alpha - 1) \le \sum_{\substack{i=1 \\ i \ne l}}^{n} (e_{li} - 1) \le (n-1)\left(\frac{1}{\alpha} - 1\right).$$

Then by dividing the above expression by $(n-1)$ and using (6.55) we obtain:

$$\alpha - 1 \leq CI(C) \leq \frac{1}{\alpha} - 1 \qquad (6.56)$$

which is the desired assertion. □

The above theorem shows, for example, that $K(C) < 0.09$ translates to $CI(C) < 0.1$, which in turns translates to a low (in most cases considered as acceptable) consistency ratio $CR(C)$. Unfortunately, this relationship cannot be reversed. One can easily give examples of matrices in which a small $CI(C)$ coexists with a relatively high $K(C)$. For example, when

$$C = \begin{pmatrix} 1 & 1 & 1 & 1 & 1 & 1 & 9 \\ 1 & 1 & 1 & 1 & 1 & 1 & 1 \\ 1 & 1 & 1 & 1 & 1 & 1 & 1 \\ 1 & 1 & 1 & 1 & 1 & 1 & 1 \\ 1 & 1 & 1 & 1 & 1 & 1 & 1 \\ 1 & 1 & 1 & 1 & 1 & 1 & 1 \\ \frac{1}{9} & 1 & 1 & 1 & 1 & 1 & 1 \end{pmatrix},$$

then $CI(C) = 0.102$, $CR(C) = 0.076$ (hence the consistency can be considered as acceptable), but $K(C) = 0.88$ (see the corner PC matrix [38]). In other words, the Koczkodaj inconsistency index is a more restrictive criterion than $CI(C)$. An interesting comparison of *Saaty's* $CI(C)$ and *Koczkodaj's* $K(C)$ can be found in the work of *Bozóki* and *Rapscák* [24].

6.3.8.3 Koczkodaj index and the principal eigenvalue

Since the definition of $CI(C)$ is based on the principal eigenvalue λ_{max}, then the Koczkodaj inconsistency index can also be used to limit the value λ_{max}. Following Theorem 8 we obtain

$$\alpha - 1 \leq \frac{\lambda_{max} - n}{n - 1} \leq \frac{1}{\alpha} - 1.$$

Hence,

$$(n-1)(\alpha-1) + n \leq \lambda_{max} \leq (n-1)\left(\frac{1}{\alpha} - 1\right) + n.$$

Because $-1 < (\alpha - 1) \leq 0$, then $-(n-1) < (n-1)(\alpha - 1)$. This means that $1 < (n-1)(\alpha-1) + n$. This is just another evidence of the well-known fact [155] that the principal eigenvalue λ_{max} of the PC matrix is positive.

6.3.9 Triad-based average inconsistency indices

In [118], *Kułakowski* and *Szybowski* proposed four inconsistency indices based on the triad inconsistency $K_{i,j,k}(C)$ (6.23). Let T be the set of all possible subsets of distinct indices in the form $\{i, j, k\}$ (set of triads) such that $i, j, k = 1, \ldots, n$. Thus, the first proposed index is

$$I_1(C) \overset{df}{=} \frac{1}{\binom{n}{3}} \sum_{\{i,j,k\} \in T} K_{ijk}(C). \tag{6.57}$$

As there are $6/(n(n-1)(n-2))$ different subsets $\{i, j, k\}$ for $i, j, k = 1, \ldots, n$, I_1 is a simple arithmetic average of all the local inconsistencies K_{ijk}. The second index is based on the same idea, but the local inconsistency is squared and the sum is square rooted, given as:

$$I_2(C) \overset{df}{=} \frac{1}{\binom{n}{3}} \sqrt{\sum_{\{i,j,k\} \in T} K_{ijk}^2}.$$

Both indices can be combined together using linear operations. Therefore, two index families have also been proposed. The first one is a combination of the Koczkodaj inconsistency index and $I_1(C)$ defined as:

$$I_\alpha(C) \overset{df}{=} \alpha K(C) + (1 - \alpha) I_1(C),$$

where $0 \leq \alpha \leq 1$, and the second one that combines together three indices: $K(C), I_1(C)$ and $I_2(C)$.

$$I_{\alpha,\beta}(C) \overset{df}{=} \alpha K(C) + \beta I_1(C) + (1 - \alpha - \beta) I_2(C).$$

6.3.10 Index of determinants

Peláez and *Lamata* [140] noticed that the determinant of the 3×3 PC matrix

$$C = \begin{pmatrix} 1 & c_{12} & c_{13} \\ 1/c_{12} & 1 & c_{23} \\ c_{13} & 1/c_{23} & 1 \end{pmatrix}$$

is given as

$$\det C = \frac{c_{13}}{c_{12}c_{23}} + \frac{c_{12}c_{23}}{c_{13}} - 2. \tag{6.58}$$

It is easy to see[5] that $\det(C) \geq 0$. By construction of the determinant (6.58) it is clear that the higher the difference between c_{13} and $c_{12}c_{23}$, i.e., the more

---

[5]To prove that, it is enough to consider $\det C = f(x) - 2$ where

$$f(x) = x + 1/x \text{ and } x = \frac{c_{13}}{c_{12}c_{23}}.$$

Since $f'(x) = 1 - 1/x^2$, i.e., $f'(1) = 0$ thus f takes the minimum for $x = 1$ where $f(1) = 2$.

inconsistent the only triad, the higher the value of $\det(C)$. In other words, for a 3×3 PC matrix its determinant can be used as an inconsistency indicator. The authors proposed to extend this reasoning to PC matrices of any size by calculating determinants for any possible 3×3 submatrices

$$T_{ijk} = \begin{pmatrix} 1 & c_{ij} & c_{ik} \\ 1/c_{ij} & 1 & c_{kj} \\ c_{ik} & 1/c_{kj} & 1 \end{pmatrix}$$

of the main PC matrix C, then compute its arithmetic mean. Therefore, the index of determinants takes the form

$$CI^*(C) \stackrel{df}{=} \frac{1}{\binom{n}{3}} \sum_{\{i,j,k\} \in T} \det T_{ijk},$$

where, similarly to before, T is a set of all possible subsets of indices[6] $\{i, j, k\}$ such that $i, j, k = 1, \ldots, n$.

The index of determinants is slightly similar to I_1. The simple expression $K_{ijk}(C)$ in I_1 has been replaced by the more complex $\det T_{ijk}$ in CI^*. The Koczkodaj triad inconsistency index can also be used for constructing the index of determinants. To see this, let us assume for a moment[7] that

$$1 - \frac{c_{ij}}{c_{ik}c_{kj}} = K_{ijk}(C).$$

If it is so, then

$$\frac{c_{ij}}{c_{ik}c_{kj}} = 1 - K_{ijk}(C) \quad \text{and} \quad \frac{c_{ik}c_{kj}}{c_{ij}} = \frac{1}{1 - K_{ijk}(C)}.$$

Therefore, the determinant of T_{ijk} can be written as

$$\det T_{ijk} = \frac{1}{1 - K_{ijk}(C)} - (1 - K_{ijk}(C)).$$

Thus, providing that $g(x) = \frac{1}{1-x} - (1 - x)$, the index of determinants can be written as

$$CI^*(C) = \frac{1}{\binom{n}{3}} \sum_{\{i,j,k\} \in T} g \circ K_{ijk}(C). \tag{6.59}$$

Comparing both (6.57) and (6.59) their similarity becomes clear. The only difference is that in the case of CI^* the local inconsistency $K_{ijk}(C)$ is transformed by the monotonically growing on $[0, 1]$ function g.

---

[6] Note that when C is reciprocal then the value $\det T_{ijk}$ does not depend on the order of the indices i, j, k, i.e., $\det T_{ijk} = \det T_{jik} = \ldots$, etc. The same applies to $K_{ijk}(C)$.

[7] The same result we get when assuming $1 - \frac{c_{ik}c_{kj}}{c_{ij}} = K_{ijk}(C)$.

6.3.11 Ambiguity index

Salo and *Hämäläinen* [172] assumed that every single comparison c_{ij} is approximated by the expressions $c_{ik}c_{kj}$ for $k = 1, \ldots, n$. The lowest and highest values of this expression determine the admissible range of values for c_{ij}. Let R be the matrix composed of the admissible intervals of values for pairwise comparisons in some $C = [c_{ij}]$.

$$
R = \begin{pmatrix}
(\underline{r}_{11}, \overline{r}_{11}) & \cdots & (\underline{r}_{1n}, \overline{r}_{1n}) \\
\vdots & \ddots & \vdots \\
(\underline{r}_{n1}, \overline{r}_{n1}) & \cdots & (\underline{r}_{nn}, \overline{r}_{nn})
\end{pmatrix},
$$

where

$$
\underline{r}_{ij} = \min\{c_{ik}c_{kj} \mid k = 1, \ldots, n\}
$$

and

$$
\overline{r}_{ij} = \max\{c_{ik}c_{kj} \mid k = 1, \ldots, n\}.
$$

The distance between $\underline{r}_{ij}$ and $\overline{r}_{ij}$ has the meaning of an inconsistency indicator. Indeed, one may expect that the more consistent C is, the smaller the distance between the minimal and maximal approximations of a single comparison. This observation allowed the researchers to formulate a new inconsistency index $I_{SH}(C)$.

$$
I_{SH}(C) \overset{df}{=} \frac{1}{\binom{n}{2}} \sum_{i=1}^{n-1} \sum_{j=i+1}^{n} \frac{\overline{r}_{ij} - \underline{r}_{ij}}{(1 + \overline{r}_{ij})(1 + \underline{r}_{ij})}.
$$

6.3.12 Other indices

Another matrix-based inconsistency index for a multiplicative PC matrix comes from *Cavallo* and *D'Apuzzo* [8, 34]. Similarly to $K_{ijk}(C)$, it is based on the ratio $c_{ij}/c_{ik}c_{kj}$.

$$
I_{CD}(C) \overset{df}{=} \left(\prod_{\{i,j,k\} \in T} \max\left\{ \frac{c_{ij}}{c_{ik}c_{kj}}, \frac{c_{ik}c_{kj}}{c_{ij}} \right\} \right)^{\frac{1}{\binom{n}{3}}},
$$

where T is a set of all possible subsets of indices $\{i, j, k\}$ such that $i, j, k = 1, \ldots, n$. It is worth noting that for a consistent PC matrix the index I_{CD} is 1 while most other indices are 0. Of course, the higher the value of $I_{CD}(C)$ the more inconsistent C is.

An interesting proposal for an inconsistency index comes from *Stein* and *Mizzi* [181]. They proposed to sum elements in columns, i.e., $s_j = \sum_{i=1}^{n} c_{ij}$, then compute the sum of the inverse of s_j, i.e., $S(C) = \sum_{j=1}^{n} s_j^{-1}$. It can be proven that $S(C)$ is 1 if and only if C is consistent, otherwise it is smaller than 1. Based on that observation, the harmonic consistency index has been proposed as:

$$I_H(C) \overset{df}{=} \frac{n+1}{n(n-1)}\left(n/S(C) - n\right).$$

There are also other methods for determining inconsistency not mentioned in this chapter. For example, *Kou* and *Lin* proposed the Cosine Consistency Index [107, 97], *Kułakowski* et al. defined a generalized inconsistency index based on alo-groups [117], *Gass* and *Rapcsák* showed that inconsistency can be determined using the Frobenus norm between matrices [69], a cycle-based consistency index has been proposed by *Szybowski* [188], *Kazibudzki* devised a couple of matrix-based indices [94], *Ramík* and *Korviny* proposed a ranking-based inconsistency index [148], *Shiraishi* and *Obata* [176, 177] indicated that a good candidate for an inconsistency index is the third coefficient of the characteristic polynomial of C, and others. Similarities between the latter and the index of determinants are shown by *Brunelli* [27]. Further information on inconsistency indices can be found in [31, 30].

6.4 Properties of inconsistency indices

Despite the relatively large number and variety of inconsistency indices, we can try to isolate some of their common features. Some postulates concerning the features of inconsistency indices were proposed as axioms. We will briefly discuss them in this section. However, because it is not our goal to create an axiomatic system regarding inconsistency indices, we will not consider the consistency of the set of axioms, their mutual independence, etc. Most of the properties discussed here come from the work of *Brunelli* and *Fedrizzi* [33]. Some other features come from other authors, e.g., the idea of normalization was raised by *Koczkodaj* and *Urban* [106]. Further discussion on various indices can be found in the numerous papers [47, 106, 29, 104, 117, 34, 24, 32, 149].

6.4.1 The uniqueness of the consistency indication

One of the basic features of inconsistency indices is their ability to detect inconsistencies, i.e., determining whether a given matrix is consistent or inconsistent. If a matrix is consistent, the given index takes some value $\alpha \in \mathbb{R}_+$, and when it is inconsistent the index differs from α. According to the first axiom proposed by *Brunelli* and *Fedrizzi* [33], there should be exactly one such α for the given index, which means that the matrix is consistent. Of course, different indices may have different alpha values. For example, when

C is consistent, which means, according to (Def. 27), that there is no triad $\{i, j, k\}$ such that $c_{ij} \neq c_{ik}c_{kj}$, then $CI(C) = K(C) = 0$ and any other value of these indices indicates the existence of an inconsistent triad. For the same matrix, the *Cavallo* and *D'Apuzzo* index returns one, i.e., $I_{CD}(C) = 1$. It is easy to verify that for I_{CD} 1 is the only value indicating a consistent matrix.

A similar version of this property has been proposed by *Koczkodaj* and *Urban* [106]. According to their research, the index value for a consistent matrix should equal 0. This postulate is a special case of the previous proposal where α is fixed and equals 0. On the other hand, every inconsistency index that meets the uniqueness of the consistency indication can be easily transformed so that it equals 0 for a consistent matrix. An example is $I'_{CD}(C) = I_{CD}(C) - 1$.

6.4.2   Invariance under permutation of alternatives

In the pairwise comparisons method, the order of alternatives is contractual. This means that it is not important whether alternative x is labeled by 1 and y by 2. If y is preferred then it will be the winner of direct comparisons, i.e., $c_{21} = c_{yx} > 1$. For this reason, it is reasonable to expect that inconsistency indices are also insensitive to the order of alternatives. This postulate will translate into a requirement of the invariance of indices from the order of alternatives. In other words, if C' is a PC matrix obtained from C by swapping the i-th and j-th columns and i-th and j-th rows, then for both C and C' the inconsistency measured by the inconsistency index should be identical.

Example 11. *Let us consider a PC matrix:*

$$
C = \begin{pmatrix}
1 & \frac{8}{9} & \frac{2}{7} & \frac{1}{2} & \frac{5}{2} \\
\frac{9}{8} & 1 & \frac{2}{5} & \frac{5}{7} & \frac{8}{3} \\
\frac{7}{2} & \frac{5}{2} & 1 & \frac{9}{8} & 6 \\
2 & \frac{7}{5} & \frac{8}{9} & 1 & 7 \\
\frac{2}{5} & \frac{3}{8} & \frac{1}{6} & \frac{1}{7} & 1
\end{pmatrix}.
$$

A permutation matrix $P = [p_{ij}]$ is a matrix in which for every row and for every column there exists only one element equal to 1. All other elements are 0. For example, a 5×5 permutation matrix swapping the second and the fourth alternative looks as follows:

$$
P = \begin{pmatrix}
1 & 0 & 0 & 0 & 0 \\
0 & 0 & 0 & 1 & 0 \\
0 & 0 & 1 & 0 & 0 \\
0 & 1 & 0 & 0 & 0 \\
0 & 0 & 0 & 0 & 1
\end{pmatrix}.
$$

The matrix C with altered 2 and 4 alternatives is given as:

$$PCP^T = \begin{pmatrix} 1 & \frac{1}{2} & \frac{2}{7} & \frac{8}{9} & \frac{5}{2} \\ 2 & 1 & \frac{8}{9} & \frac{7}{5} & 7 \\ \frac{7}{2} & \frac{9}{8} & 1 & \frac{5}{2} & 6 \\ \frac{9}{8} & \frac{5}{7} & \frac{2}{5} & 1 & \frac{8}{3} \\ \frac{2}{5} & \frac{1}{7} & \frac{1}{6} & \frac{3}{8} & 1 \end{pmatrix}.$$

It is easy to compute that indeed $CI(C) = IC(PCP^T) = 0.01258$, $K(C) = K(PCP^T) = 0.4666$, etc.

6.4.3 Judgment intensification

This property reflects the expectation that if the views of experts intensify, the inconsistency of such views will increase rather than decrease. By judgment intensification, we mean transforming every single comparison c_{ij} through the mapping $f_b(x) = x^b$ where $b > 1$. Formally, the inconsistency index I has this property if for two matrices $C = [c_{ij}]$ and $C_b = [c_{ij}^b]$ such that $b > 1$ and it holds that $I(C) \leq I(C_b)$. This property is consistent with the observation that the larger the scale range, the greater the inconsistency of the random matrices (Fig. 6.1). Indeed, raising comparisons to power corresponds to extending the assessment scale. Many inconsistency indices have the proposed property. One of the exceptions is the harmonic consistency index [181, 33].

6.4.4 Local monotonicity

In general, the local monotonicity property meets the expectation that if we locally increase the inconsistency (we change ceteris paribus an individual comparison c_{ij}) then the inconsistency index should also increase (or at least not decrease) its value.

Let us consider the following consistent PC matrix

$$C = \begin{pmatrix} 1 & \frac{2}{3} & \frac{2}{7} & \frac{2}{5} \\ \frac{3}{2} & 1 & \frac{3}{7} & \frac{3}{5} \\ \frac{7}{2} & \frac{7}{3} & 1 & \frac{7}{5} \\ \frac{5}{2} & \frac{5}{3} & \frac{5}{7} & 1 \end{pmatrix}.$$

By changing any element of C we increase the local inconsistency. For example, changing c_{13} from $2/7$ to $3/7$ increases the distance between $c_{13} = 3/7$ and the product $c_{12}c_{13} = 2/7$. If we continue increasing c_{13} more and more, the discrepancy between c_{13} and $c_{12}c_{13}$ becomes bigger and bigger. In such a situation, it seems natural to expect that the value of the inconsistency index will increase. Such an increase can be seen for CI and the matrix C (Fig. 6.2).

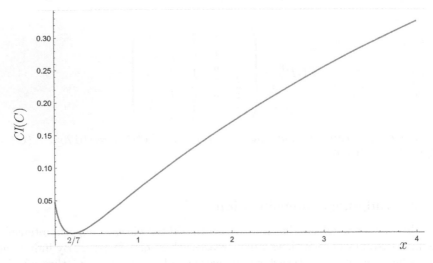

Figure 6.2: Local monotonicity for C, where $c_{13} = x, c_{31} = 1/x$

The local monotonicity property is defined differently by different researchers. *Brunelli* and *Fedrizzi* [33] propose changes in the selected elements c_{ij} and c_{ji} by raising them to the power. *Koczkodaj* and *Urban* [106] define the local monotonicity with respect to the triad and the local inconsistency index. In this approach, the increase in distance between c_{ij} and $c_{ik}c_{kj}$ needs to be accompanied by the increase in the local inconsistency index based on the given triad $\{a_i, a_j, a_k\}$.

6.4.5   Continuity

The idea of continuity meets the expectation that each, even the smallest, change in comparisons will translate into a corresponding change in inconsistency. So if we ask experts to correct their opinion regarding a specific comparison, then it seems natural to expect that we immediately know whether such a change has increased or decreased the inconsistency. Continuity seems to be particularly important in mathematical modeling, and where minimization of inconsistencies is involved. Many indices are continuous, but some are not. *Brunelli* proves that, for example, *Barzilai's* relative error is not continuous [33].

6.4.6   Normalization

The values of consistency indices are not expressed in any specific units. They have no physical interpretation either. Hence, it is difficult to assess whether a given inconsistency value is high or low. It is also difficult to compare them with each other. Therefore, *Koczkodaj* and *Urban* [106] propose that all indices should be normalized, i.e., their values should be in the range $[0,1]$. Most of the indices that can be found in the literature do not meet this requirement. An exception is the *Koczkodaj* inconsistency index as $\min\{1 - x/y, 1 - y/x\}$ for $x, y \in \mathbb{R}_+$ is always smaller than 0. However, many of them can be scaled to meet this condition. For example,

$$\widehat{CI}(C) = \frac{2q\,CI(C)}{(q-1)^2},$$

where q is the range of a scale is the normalized version of Saaty's consistency index. In particular, the normalized consistency index for the fundamental scale may look as follows:

$$\widehat{CI}(C) = \frac{9}{32}\,CI(C).$$

6.4.7 Contraction

Koczkodaj and *Urban* [106] also claim that for two PC matrices C and C' where C' is the sub-matrix of C the inconsistency measured by the given index I for C cannot be smaller than the inconsistency for C', i.e., $I(C') \leq I(C)$. In other words, adding alternatives (and their comparisons) to the matrix cannot reduce the amount of inconsistency in the matrix. Obviously, most indices do not have this property, especially those that average local inconsistencies. In their case, adding new consistent comparisons will reduce the overall inconsistency. Nevertheless, it was shown [111, p. 117] that the *Koczkodaj* index has this property. Indeed, the larger matrix has more triads than its submatrix. For this reason, the maximally inconsistent triad in a subset cannot be greater than the maximally inconsistent triad in the whole set.

Chapter 7

Inconsistency of Incomplete PC Matrices

7.1 Introduction

In the previous chapter, we dealt with ways of determining the level of inconsistency for PC matrices. However, all the matrices were complete, i.e., for every pair of alternatives, there was a certain real number corresponding to their relative priority. This is not the case for incomplete PC matrices (Section 2.3). It is possible that some alternatives do not have their direct comparison. So there is a possibility that the consistency condition $c_{ij} = c_{ik}c_{kj}$ (see 6.2), which was the starting point for considering inconsistency, in some cases does not make sense. Indeed, if any of the values c_{ij}, c_{ik} or c_{kj} do not exist, we cannot decide whether the triad a_i, a_k, a_j is consistent or not. For this reason, it will be necessary to extend the triad-based definition (Def. 27) of the concept of inconsistency. Such a new definition will allow, on the one hand, a better description of the phenomenon of inconsistency; on the other hand, it will enable an effective extension of selected inconsistency indices from Chapter 6 to include incomplete PC matrices.

7.2   Preliminaries

Similarly as for complete PC matrices (Section 6.3), the inconsistency indices for incomplete PC matrices can be divided into two groups. Those for which we need to know the ranking of alternatives first, and those for which we only need the PC matrix elements (and no ranking along the way is calculated). This division allows, in particular, two strategies to be formulated for extending existing indices to incomplete matrices. According to the first one, the ranking is calculated for each of the considered matrices using the method for an incomplete ranking matrix. According to the second, local inconsistency indices are calculated for some selected groups of elements, which are then aggregated to obtain the final result. This division is, of course, arbitrary. In this chapter, we will also discuss indices that elude such a simple classification.

7.2.1   Matrix-based indices

By matrix-based indices we mean those which use matrix elements directly (Section 6.3). In the case of complete PC matrices, they used the notion of a triad. But, as mentioned before, using triads in the context of incomplete PC matrices is quite risky. What if the triad does not exist? The first thought that comes to mind is: let us skip non-existent triads and try to use those that exist. However, it soon turns out that this is not the right strategy. Let us consider the following example [119]:

$$
C = \begin{pmatrix}
1 & 1/2 & ? & ? & ? & ? & 1/7 \\
2 & 1 & ? & 6 & 4 & 2 & ? \\
? & ? & 1 & 4 & 3 & 3/2 & ? \\
? & 1/6 & 1/4 & 1 & ? & ? & 1/2 \\
? & 1/4 & 1/3 & ? & 1 & ? & 1/4 \\
? & 1/2 & 2/3 & ? & ? & 1 & 1/3 \\
7 & ? & ? & 2 & 4 & 3 & 1
\end{pmatrix}. \tag{7.1}
$$

It is easy to see that the undirected graph P_C of (7.1) does not contain cycles shorter than 4 (Fig 7.1). For this reason, there are no triads in C and yet C is irreducible (P_C is connected, see Def. 7), i.e., one can calculate the ranking based on it.

The above example shows that the notion of a triad is insufficient to define inconsistency in the context of incomplete PC matrices. To remedy this, we will use the concept of a cycle in the new, extended definition of inconsistency. It is worth noting that the idea of using cycles for determining inconsistency is not new. One can find it, for example, in [104, 188, 25, 119].

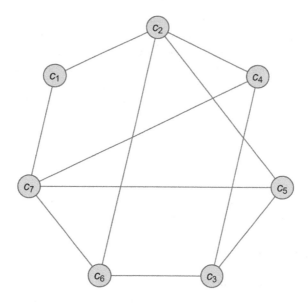

Figure 7.1: Undirected graph of the irreducible matrix C containing no cycles shorter than 4

Definition 31. *A PC matrix $C = [c_{ij}]$ is said to be inconsistent if there is a cycle $a_{i_1}, \ldots, a_{i_m}$ in P_C such that $c_{i_1 i_2} c_{i_2 i_3} \cdot \ldots \cdot c_{i_{m-1} i_m} \neq c_{i_1 i_m}$. Otherwise C is consistent.*

For complete PC matrices, the extended definition of inconsistency (Def. 31) and its triad-based counterpart (Def. 27) are equivalent. To find out, let us consider the following theorem [119]:

Theorem 9. *Every complete PC matrix C is inconsistent according to (Def. 27) if and only if it is inconsistent according to (Def. 31)*

Proof. "$\Rightarrow$" Let C be inconsistent according to (Def. 27), i.e., there is a triad such that $c_{ik} c_{kj} \neq c_{ij}$. As the triad is also a cycle, then C is also inconsistent according to (Def. 31).

"$\Leftarrow$" Let C be inconsistent according to (Def. 31), i.e., there is a cycle $s = a_{i_1}, \ldots, a_{i_m}$ such that $c_{i_1 i_2} c_{i_2 i_3} \cdot \ldots \cdot c_{i_{m-1} i_m} \neq c_{i_1 i_m}$.

If C is consistent according to (Def. 27) then every triad must be consistent. In particular, it also holds that $c_{i_1 i_2} c_{i_2 i_3} = c_{i_1 i_3}$. Thus, we can rewrite the assumption as: $c_{i_1 i_3} \cdot \ldots \cdot c_{i_{m-1} i_m} \neq c_{i_1 i_m}$. Similarly $c_{i_1 i_3} c_{i_3 i_4} = c_{i_1 i_4}$, which implies that $c_{i_1 i_4} \cdot \ldots \cdot c_{i_{m-1} i_m} \neq c_{i_1 i_m}$. Repeating this reasoning at most m times we subsequently obtain the inequalities $c_{i_1 i_5} \cdot \ldots \cdot c_{i_{m-1} i_m} \neq c_{i_1 i_m}$, $c_{i_1 i_6} \cdot \ldots \cdot c_{i_{m-1} i_m} \neq c_{i_1 i_m}$, and finally $c_{i_1 i_{m-1}} \cdot c_{i_{m-1} i_m} \neq c_{i_1 i_m}$. According to the assumption, however, every triad is consistent. Contradiction. $\qquad \square$

In general, for incomplete PC matrices, the first definition (Def. 27) does not apply. In particular, it is possible that there are no triads in the given PC matrix (so the condition $c_{ik}c_{kj} = c_{ij}$ for every distinct $i, j, k = 1, \ldots, n$ is satisfied), while for longer cycles this equality may not occur. However, if we would like to use Definition 27 in the context of incomplete PC matrices, we could safely say that if a PC matrix C is inconsistent according to (Def. 27) then it is also inconsistent with respect to (Def. 31).

For the purpose of further consideration, let us denote the ratio of R_s [119, 104] as

$$R_s \stackrel{df}{=} \frac{c_{i_1 i_2} \cdot \ldots \cdot c_{i_{m-1} i_m}}{c_{i_1 i_m}}, \tag{7.2}$$

where $s = a_{i_1}, \ldots, a_{i_m}$ is a cycle in a graph of C. R_s denotes the local inconsistency[1] connected with the cycle s.

7.2.2   Ranking-based indices

Some inconsistency indices require a ranking to be created. The method of their construction is most often based on the observation that the ratio of priorities of two alternatives should ideally correspond to the value of their direct comparison[2] (6.1), however, the more inconsistent the matrix, the more average the difference between the ratio $w(a_i)/w(a_j)$ and c_{ij}. This regularity also works for incomplete PC matrices. In this case, however, we have to use priority ranking methods for incomplete PC matrices. Fortunately, both the most commonly used priority-deriving methods EVM and GMM have their incomplete counterparts (Chapter 4). Therefore, the ranking-based indices for incomplete PC matrices will use them instead of the original EVM or GMM. Both incomplete EVM and GMM extensions adopt the idea that the unknown comparisons $c_{ij} =?$ are identified with the ideal ratios $w(a_i)/w(a_j)$. As a result, the local inconsistency for unknown comparisons is 0, hence, the missing comparisons do not contribute to the inconsistency of an incomplete PC matrix. This observation allows us to simplify some formulas.

7.3   Quantifying inconsistency

7.3.1   Saaty–Harker consistency index

The inconsistency index defined by *Saaty* (Section 6.3.1) can also be extended to incomplete PC matrices [76, 119]. The approach proposed by *Harker* [76] comes in handy here. The difference between *Saaty's* original proposal and

---

[1]It is worth noting that R_s is not a special case of the cycle index introduced by Szybowski [188] as $c_{i_1 i_2} \cdot \ldots \cdot c_{i_{m-1} i_m} = c_{i_1 i_m}$ implies $R_s = 1$.

[2]We have even proved that this is the case for a consistent PC matrix (Theorem 4).

Harker's solution lies in the matrix for which the spectral radius is calculated. Saaty uses the PC matrix itself, while *Harker* replaces it with the auxiliary matrix B (4.4). Thus, the *Saaty–Harker* index can be defined as:

$$\widetilde{CI} = \frac{\tilde{\lambda}_{max} - n}{n - 1},$$

where $\tilde{\lambda}_{max}$ is the principal eigenvalue of B (4.4). *Harker* [76, p. 356] showed that $\widetilde{CI}$ can also be written as:

$$\widetilde{CI} = -1 + \frac{1}{n(n-1)} \left(\sum_{i=1}^{n} m_i + \sum_{\substack{1 \le i < j \le n \\ c_{ij} \neq ?}} \left(c_{ij} \frac{\tilde{w}(a_j)}{\tilde{w}(a_i)} + c_{ji} \frac{\tilde{w}(a_i)}{\tilde{w}(a_j)} \right) \right),$$

where $\tilde{w} = [\tilde{w}(a_1), \dots, \tilde{w}(a_n)]^T$ is the principal eigenvector of B. *Harker* also proved that if B is consistent then $\tilde{\lambda}_{max} = n$, and as follows $\widetilde{CI} = 0$.

7.3.2 Logarithmic least squares criterion

In (Section 6.3.6) we can see that the logarithmic least squares criterion can also be used as an inconsistency measure. Indeed, the more consistent the PC matrix C, the smaller the error

$$\mathcal{E}(C) \overset{df}{=} \sum_{i,j=1}^{n} \left(c_{ij} - \frac{w(a_i)}{w(a_j)} \right)^2,$$

where w is the priority vector computed using LLSM (Section 3.3.3). Of course, the incompleteness of C requires some correction of the above formula. Fortunately, in this case it is enough not to take into account only those components of the sum for which there are no comparisons. Thus, the error formula takes the form:

$$\widehat{\mathcal{E}}(C) \overset{df}{=} \sum_{\substack{i,j=1 \\ c_{ij} \neq ?}}^{n} \left(c_{ij} - \frac{w(a_i)}{w(a_j)} \right)^2.$$

Thus, the value of the inconsistency index [23] can be computed as

$$\widehat{L}(C) = \min_{w} \widehat{\mathcal{E}}(C),$$

where w is the real and positive vector in $\mathbb{R}^n$.

7.3.3   Oliva–Setola–Scala inconsistency index

An interesting idea for measuring the inconsistency of incomplete PC matrices comes from *Oliva*, *Setola*, and *Scala* [137]. As the inconsistency index, they propose the use of

$$O(C) = \rho(D^{-1}S) - 1,$$

where ρ means the spectral radius, D is the degree matrix of P_C (Def. 9) and $S = \tilde{C} - Id$. The matrix $\tilde{C}$ is obtained from C by replacing every missing comparison by 0.

7.3.4   Koczkodaj inconsistency index

The Koczkodaj index [98, 104] spots the highest local triad's inconsistency. It is a typical representative of matrix-based indices. Because the notion of a triad is not enough to define the local inconsistency for incomplete matrices (Section 7.2.1), triads have to be replaced by cycles.

Let us consider an irreducible and incomplete PC matrix C and its graph T_C. Then, let the inconsistency of a cycle[3] longer than[4] 2, i.e., $s \in \mathcal{S}_{C,2}$ (see. Def. 14) be defined as:

$$K_s \stackrel{df}{=} \min\left\{|1 - R_s|, |1 - R_s^{-1}|\right\}. \tag{7.3}$$

The Koczkodaj index for the incomplete PC matrix C is defined as [119]:

$$\tilde{K} \stackrel{df}{=} \begin{cases} \max\{K_s : s \in \mathcal{S}_{C,2}\} & |\mathcal{S}_{C,2}| > 0 \\ 0 & |\mathcal{S}_{C,2}| = 0 \end{cases}.$$

When $|\mathcal{S}_{C,2}| = 0$ then the $n \times n$ matrix C is irreducible but contains too few comparisons (exactly $n - 1$) to be inconsistent.

7.3.5   Triad-based average inconsistency indices

In [118] *Kułakowski* and *Szybowski* defined a bunch of matrix-based inconsistency indicators (Section 6.3.9). These indices intensively use the concept of the local inconsistency introduced by *Koczkodaj* [104]. However, they can easily be extended to include incomplete PC matrices. It is enough to replace triad inconsistency by cycle inconsistency. Hence, providing that the PC matrix C is irreducible and incomplete, we may define:

$$\tilde{I}_1 \stackrel{df}{=} \begin{cases} \dfrac{\sum_{s \in \mathcal{S}_{C,2}} K_s}{|\mathcal{S}_{C,2}|} & |\mathcal{S}_{C,2}| > 0 \\ 0 & |\mathcal{S}_{C,2}| = 0 \end{cases},$$

---

[3]It is worth noting that if $s = a_i a_k a_j$ then $K_s = K_{i,k,j}$ (7.3).

[4]Cycles with the length 2 are always consistent as $c_{ij}c_{ji}/c_{ii} = 1$, thus they are not relevant from the point of inconsistency of C.

and,

$$\widetilde{I}_2 \overset{df}{=} \begin{cases} \frac{\sqrt{\sum_{s \in \mathcal{S}_{C,2}} K_s^2}}{|\mathcal{S}_{C,2}|} & |\mathcal{S}_{C,2}| > 0 \\ 0 & |\mathcal{S}_{C,2}| = 0 \end{cases}.$$

In addition, I_α and $I_{\alpha,\beta}$ indices can be easily extended to cover incomplete PC matrices:

$$\widetilde{I}_\alpha \overset{df}{=} \alpha \widetilde{K} + (1 - \alpha)\widetilde{I}_1,$$

$$\widetilde{I}_{\alpha,\beta} \overset{df}{=} \alpha \widetilde{K} + \beta \widetilde{I}_1 + (1 - \alpha - \beta)\widetilde{I}_2.$$

7.3.6 Ambiguity index

The idea of triad inconsistency (6.2, 6.23) is based on the observation that every product $c_{ik}c_{kj}$ for any $k = 1, \ldots, n$ approximates c_{ij}. If C is consistent then $c_{ik}c_{kj} = c_{ij}$. If not $c_{ik}c_{kj} \approx c_{ij}$. The distance between these two expressions determines the level of local inconsistency within the matrix C. The Salo-Hämäläinen index (also called ambiguity index) measures the average differences between the highest and lowest approximations $c_{ik}c_{kj}$ for different comparisons c_{ij}.

When C is incomplete (but irreducible) it is possible, however, that for certain c_{ij} there will be no triads containing it, i.e., there will be no approximations in the form of $c_{ik}c_{kj}$. Fortunately, irreducibility implies that between every two vertices of P_C there is a path $p = a_q, a_{i_2}, a_{i_2}, \ldots, a_{i_{m-1}}, a_j$, i.e., there are appropriate comparisons: $c_{q,i_2}, c_{i_2 i_3}, \ldots, c_{i_{m-1},j}$. Let the product induced by p be denoted as $\pi_p = c_{q,i_2} c_{i_2 i_3} \cdot \ldots \cdot c_{i_{m-1},j}$. It is easy to observe that π_p is also a good approximation of c_{ij}. Following [119], let us define:

$$\underset{\sim}{r}_{ij} \overset{df}{=} \min \{\pi_p \mid p \in \mathcal{P}_{C,i,j}\}$$

and

$$\widetilde{r}_{ij} \overset{df}{=} \max \{\pi_p \mid p \in \mathcal{P}_{C,i,j}\}.$$

It is worth noting that, even for an incomplete PC matrix, it holds that $\underset{\sim}{r}_{ij}, \widetilde{r}_{ij} \neq \emptyset$. Thus, the modified Salo-Hamalainen index can be defined as:

$$\widetilde{I}_{SH} \overset{df}{=} \frac{2}{n(n-1)} \sum_{i=1}^{n-1} \sum_{j=i+1}^{n} \frac{\widetilde{r}_{ij} - \underset{\sim}{r}_{ij}}{(1 + \widetilde{r}_{ij})\left(1 + \underset{\sim}{r}_{ij}\right)}.$$

7.3.7 Geometric consistency index

The geometric consistency index (Section 6.3.3) measures the average squared logarithm of error:

$$e_{ij} = c_{ij} \frac{w(a_j)}{w(a_i)},$$

where $i, j = 1, \ldots, n$, and w is the priority vector calculated using GMM (Section 3.2). For incomplete PC matrices, the priority vector w needs to be calculated using an appropriate incomplete version of GMM (Section 4.2). Let us denote

$$\widehat{e}_{ij} = c_{ij} \frac{\widehat{w}(a_j)}{\widehat{w}(a_i)},$$

where $\widehat{w}$ is the priority vector for an incomplete PC matrix. Since this ranking method treats every missing comparison as it would be perfectly consistent, i.e., $c_{ij} = \widehat{w}(a_i)/\widehat{w}(a_j)$, we may thus assume that $\widehat{e}_{ij}$ for $c_{ij} =?$ is 1, i.e., $\log \widehat{e}_{ij} = 0$. This leads to the first version of the geometric consistency index [119] for an incomplete PC matrix defined as:

$$\widetilde{I}_{G1} \stackrel{df}{=} \frac{2}{(n-1)(n-2)} \sum_{\widehat{e} \in \mathscr{E}}^{n} \log^2 \widehat{e}_{ij},$$

where

$$\mathscr{E} \stackrel{df}{=} \left\{ e_{ij} = c_{ij} \frac{\widetilde{w}(a_j)}{\widetilde{w}(a_i)} \; : \; c_{ij} \neq ? \text{ and } i < j \right\}.$$

The above solution, however, involves the risk that the index may be low with a large number of undefined comparisons, even if the error for the other (several) comparisons is high. Therefore, not to lose the sensitivity of the index, we simply skip every e_{ij} for which $c_{ij} =?$. This leads to the modified version of the index:

$$\widetilde{I}_{G2} \stackrel{df}{=} \frac{1}{|\mathscr{E}|} \sum_{\widehat{e} \in \mathscr{E}} \log^2 \widehat{e}.$$

In the case of $\widetilde{I}_{G2}$ undefined comparisons do not contribute directly to the overall value of inconsistency.

7.3.8 Golden–Wang index

One may prove that for a consistent PC matrix $C = [c_{ij}]$ it holds that

$$\frac{c_{ik}}{c_{jk}} = \frac{w(a_i)}{w(a_j)}, \quad \text{for } k = 1, \ldots, n. \tag{7.4}$$

In other words, every single column is the ranking itself. In particular, if we rescale every column of C so that their elements sum up to 1 then $c_{ij}^* = w(a_i)$

where $C^* = [c_{ij}^*]$ is the PC matrix with the rescaled columns. If C is inconsistent, then the difference between c_{ij}^* and $w(a_i)$ can be used to determine the level of inconsistency. The average of all these distances creates the Golden–Wang inconsistency index (Section 6.3.4). Although the priority vector w can be calculated using any priority-deriving method, the authors use GMM (Section 3.2). The original Golden–Wang index can also be extended to cover incomplete PC matrices. For this purpose, however, it is not enough to replace w with its incomplete counterpart $\widehat{w}$. We also need to properly compare (possibly incomplete) columns with the resulting ranking vector.

Let us consider the k-th column of an incomplete but consistent PC matrix C and the ranking vector w [119]. For the sake of simplicity, let us assume that c_{ij} is either 1 or is undefined. Since C is consistent and all alternatives have the same priority, then the ranking vector is composed of the same $1/n$ values. For example:

$$
\begin{bmatrix} c_{1k} \\ c_{2k} \\ c_{3k} \\ c_{4k} \\ c_{5k} \end{bmatrix} = \begin{bmatrix} 1 \\ 1 \\ 1 \\ ? \\ ? \end{bmatrix}, \quad \begin{bmatrix} w(a_1) \\ w(a_2) \\ w(a_3) \\ w(a_4) \\ w(a_5) \end{bmatrix} = \begin{bmatrix} 1/5 \\ 1/5 \\ 1/5 \\ 1/5 \\ 1/5 \end{bmatrix},
$$

where $k = 1, 2$ or 3. After scaling, however, the k-th column is:

$$
\begin{bmatrix} c_{1k} \\ c_{2k} \\ c_{3k} \\ c_{4k} \\ c_{5k} \end{bmatrix} = \begin{bmatrix} 1/3 \\ 1/3 \\ 1/3 \\ ? \\ ? \end{bmatrix}.
$$

as during the scaling, undefined elements have to be ignored. Of course, $c_{ik}^* \neq w(a_i)$ for $i = 1, 2, 3$ as $1/3 \neq 1/5$. To remedy this, we must limit the priority vector to those values to which the elements defined in the given column correspond. Thus,

$$
\begin{bmatrix} c_{1k} \\ c_{2k} \\ c_{3k} \\ c_{4k} \\ c_{5k} \end{bmatrix} = \begin{bmatrix} 1 \\ 1 \\ 1 \\ ? \\ ? \end{bmatrix}, \quad \begin{bmatrix} w_k(a_1) \\ w_k(a_2) \\ w_k(a_3) \\ w_k(a_4) \\ w_k(a_5) \end{bmatrix} = \begin{bmatrix} 1/5 \\ 1/5 \\ 1/5 \\ ? \\ ? \end{bmatrix},
$$

and after scaling

$$
\begin{bmatrix} c_{1k} \\ c_{2k} \\ c_{3k} \\ c_{4k} \\ c_{5k} \end{bmatrix} = \begin{bmatrix} 1/3 \\ 1/3 \\ 1/3 \\ ? \\ ? \end{bmatrix}, \quad \begin{bmatrix} w_k(a_1) \\ w_k(a_2) \\ w_k(a_3) \\ w_k(a_4) \\ w_k(a_5) \end{bmatrix} = \begin{bmatrix} \frac{1/5}{3/5} \\ \frac{1/5}{3/5} \\ \frac{1/5}{3/5} \\ ? \\ ? \end{bmatrix} = \begin{bmatrix} 1/3 \\ 1/3 \\ 1/3 \\ ? \\ ? \end{bmatrix}.
$$

In this case, indeed $c_{ik}^* = w_k^*(a_i) = 1/3$ for $i = 1, 2, 3$. The above example shows how the Golden–Wang index can be extended. Let us define this method formally. Let C be an irreducible, incomplete PC matrix, and w be the ranking vector calculated using GMM for incomplete PC matrices. Let $\Omega = [\omega_{ij}]$ be an $n \times n$ matrix where

$$\omega_{ij} \overset{df}{=} \begin{cases} w(a_i) & \text{if } c_{ij} \neq ? \\ ? & \text{if } c_{ij} = ? \end{cases}.$$

Next, let every column in Ω and in C be scaled so that all their elements sum up to one (in both cases, undefined values are ignored). After scaling, we obtain two matrices $C^* = [c_{ij}^*]$ and $\Omega^* = [\omega_{ij}^*]$. The absolute differences between their entries form the Golden–Wang index $\widetilde{GW}$ for incomplete PC matrices, i.e.,

$$\widetilde{GW} \overset{df}{=} \frac{1}{n} \sum_{i=1}^{n} \sum_{j=1}^{n} | c_{ij}^* - \omega_{ij}^* |.$$

7.3.9   Relative error

Although *Barzilai's* Relative Error RE has been designed for additive and complete PC matrices, it can also be used for multiplicative and incomplete matrices. However, to define such an extension, we need to look at the analytical formula of this index after applying logarithmic transformation. Let $\widehat{C} = [\widehat{c}_{ij}] = [\log c_{ij}]$ be the logarithmized version of C in which every entry c_{ij} is replaced by its logarithm. This, in particular, entails the redefinition of Δ_i,

$$\Delta_i = \frac{1}{n} \sum_{j=1}^{n} \log c_{ij},$$

i.e.,

$$\Delta_i = \frac{1}{n} \log \prod_{j=1}^{n} c_{ij},$$

$$\Delta_i = \log \left(\prod_{k=1}^{n} c_{ik} \right)^{\frac{1}{n}}.$$

Thus, the matrix $X = [x_{ij}]$ where $x_{ij} = \Delta_i - \Delta_j$ obtains the form

$$X = [x_{ij}] = \left[\log \left(\prod_{k=1}^{n} c_{ik} \right)^{\frac{1}{n}} - \log \left(\prod_{k=1}^{n} c_{jk} \right)^{\frac{1}{n}} \right],$$

where,

$$x_{ij} = \log \left[\left(\prod_{k=1}^{n} c_{ik} \right)^{\frac{1}{n}} \bigg/ \left(\prod_{k=1}^{n} c_{jk} \right)^{\frac{1}{n}} \right] = \log \left[\left(\prod_{k=1}^{n} \frac{c_{ik}}{c_{jk}} \right)^{\frac{1}{n}} \right].$$

It is easy to observe that x_{ij} becomes the ratio of priorities of the i-th and j-th alternatives providing that the ranking vector is calculated using GMM (Section 3.2). Hence,

$$x_{ij} = \log \frac{\left(\prod_{k=1}^{n} c_{ik}\right)^{\frac{1}{n}}}{\left(\prod_{k=1}^{n} c_{jk}\right)^{\frac{1}{n}}} = \log \frac{w_{GMM}(a_i)}{w_{GMM}(a_j)}.$$

Let us compute the error matrix

$$E = [e_{ij}] = [\widehat{c}_{ij} - x_{ij}].$$

Their entries obtain the form

$$e_{ij} = \left[\log c_{ij} - \log \left(\prod_{k=1}^{n} \frac{c_{ik}}{c_{jk}}\right)^{\frac{1}{n}}\right],$$

i.e.,

$$e_{ij} = \left[\log c_{ij} - \log \frac{\left(\prod_{k=1}^{n} c_{ik}\right)^{\frac{1}{n}}}{\left(\prod_{k=1}^{n} c_{jk}\right)^{\frac{1}{n}}}\right].$$

Once again, we may use the GMM priority vector property (3.10):

$$e_{ij} = \left[\log c_{ij} - \log \frac{w_{GMM}(a_i)}{w_{GMM}(a_j)}\right].$$

Finally, the relative error for the logarithmized PC matrix is given as

$$RE(\widehat{C}) = \frac{\sum_{ij} e_{ij}^2}{\sum_{ij} \widehat{c}_{ij}^2}.$$

Due to the logarithmic transformation, we obtain

$$RE(C) = \frac{\sum_{ij} \left(\log c_{ij} - \log \frac{w_{GMM}(a_i)}{w_{GMM}(a_j)}\right)^2}{\sum_{ij} \log^2 c_{ij}},$$

hence

$$RE(C) = \frac{\sum_{ij} \log^2 c_{ij} \frac{w_{GMM}(a_j)}{w_{GMM}(a_i)}}{\sum_{ij} \log^2 c_{ij}}.$$

Similarly as in the other ranking-based inconsistency indices, let us replace w_{GMM} by its incomplete version w_{IGMM} (Section 4.2) and ignore all parts of the expression containing undefined comparisons. As a result of this operation, we obtain the expression:

$$\widetilde{RE}(C) \stackrel{df}{=} \frac{\sum_{c_{ij} \neq ?} \left[\log c_{ij} \frac{w_{IGMM}(a_j)}{w_{IGMM}(a_i)}\right]^2}{\sum_{c_{ij} \neq ?} \log^2 c_{ij}}.$$

The above expression works properly for every PC matrix except one that contains only elements equal 1 or ?. Of course such a matrix is consistent then we should adopt that inconsistency is 0, i.e.,

$$\widetilde{RE}(C) \overset{df}{=} 0,$$

where $C = [c_{ij}]$ and $c_{ij} \in \{1, ?\}$. This and an alternative form of this index can be found in *Kułakowski* and *Talaga* [119].

<hr>

7.4   Incompleteness and inconsistency

Does the incompleteness affect inconsistency? One may expect that, if at all, this relationship is not too strong. Since, the pairwise comparisons comes from experts, then the measured inconsistency reflects the internal inconsistency of the experts . If they are internally inconsistent then, regardless of the number of comparisons made, there must be noticeable errors and inaccuracies. Conversely, if someone has a clear opinion about a given set of alternatives, then the consistency of their answers should translate into both: complete and incomplete PC matrices. Despite the diligence of experts, they may make mistakes in the assessment. In other words, some of the errors, i.e., some of the local inconsistencies, can be caused by simple mistakes. In such cases, information redundancy in a PC matrix may help. The more pairwise comparisons there are, the greater the chance that an error will be "repaired" by other correct answers. Therefore, one can ask the question about which of the defined indices for an incomplete PC matrix is best able to mitigate the effects of errors made.

Kułakowski and *Talaga* [119] conducted an experiment in which they tested the robustness of different inconsistency indices. They created 1000 complete and consistent 7×7 PC matrices, then they disturbed them to obtain 30,000 matrices with different levels of inconsistency. Each of these matrices was then subject to the process of removing random comparisons. Finally, they got 480,000 matrices with different levels of inconsistency and incompleteness. In this set, for each incomplete matrix there was also its complete counterpart. Thus, they could compare the change in the inconsistency index value depending on the number of missing comparisons in a matrix with some fixed inconsistency. They defined the rescaled ordered distance between matrices as

$$\Delta_I(C, C_k) \overset{df}{=} \begin{cases} \frac{I(C) - I(C_k)}{\max\{I(C), I(C_k)\}} & \max\{I(C), I(C_k)\} > 0 \\ 0 & I(C) = I(C_k) = 0 \end{cases},$$

Pos.	Notation	Name	$\mathscr{D}(I)$
1.	$\widetilde{I}_1$	Cycle-based index 1	1.3368
2.	$\widetilde{I}_\alpha$	α-index, for $\alpha = 0.5$	1.5896
3.	$\widetilde{K}$	Koczkodaj index	1.692
4.	$\widetilde{I}_{\alpha,\beta}$	α, β-index, for $\alpha = \beta = 0.3$	1.7321
5.	$\widetilde{I}_{SH}$	Salo-Hamalainen index	4.1286
6.	$\widetilde{RE}$	Barzilai's relative error index	4.6748
7.	$\widetilde{I}_{G2}$	Geometric consistency index 2	4.9048
8.	O	Oliva–Setola–Scala index	4.9062
9.	$\widetilde{GW}$	Golden–Wang index	4.9359
10.	$\widetilde{L}$	Logarithmic least square condition	7.1425
12.	$\widetilde{CI}$	Saaty consistency index	7.5028
13.	$\widetilde{I}_{G1}$	Geometric consistency index 1	7.7333
14.	$\widetilde{I}_2$	Cycle-based index v. II	10.112

Table 7.1: Value of total index value differences $\mathscr{D}$ for complete and incomplete matrices after rescaling [119]. The lower the value, the better.

where I means the examined inconsistency index, C is complete (but a possibly disturbed matrix), and C_k is an incomplete PC matrix formed from C by removing k comparisons. The average value of Δ_I for matrices with k removing comparisons is denoted as:

$$D(I, k) \stackrel{df}{=} \frac{1}{|\mathcal{C}|} \sum_{C \in \mathcal{C}} \Delta_I(C, C_k), \tag{7.5}$$

and the overall results for the considered index are given as[5]:

$$\mathscr{D}(I) \stackrel{df}{=} \sum_{k=0}^{15} |D(I, k)|.$$

They obtained the following results:

The above results (Table 7.1) suggest that the cycle-based inconsistency indices for incomplete PC matrices are relatively resistant to removing alternatives. The ranking-based indices perform noticeably worse. On the other hand, the latter are easier to calculate and do not require tedious finding of all cycles in the matrix graph. Looking at the results of the experiment [119],

---

[5]In 7×7 reciprocal PC matrices there are 21 unique comparisons. The smallest number of comparisons guaranteeing the connectivity of the matrix graph is 6. Thus, we may consider matrices on 15 different levels of incompleteness, i.e., with $7, 8, \ldots, 21$ missing comparisons.

it is easy to see that, for a small number of missing comparisons, most indices perform quite well. Hence, any solution can be used to assess the inconsistency of a matrix in which only a few comparisons are missing. However, for highly incomplete matrices, solutions based on cycle analysis will probably be a better choice.

Chapter 8

Group Decisions

Although all the pairwise comparisons within a given AHP decision model can be made by one person, it does not mean, however, that AHP is a technique supporting only the decisions of individuals. On the contrary, in practice more experts are very often involved in the decision-making process. The participation of many experts allows a more accurate assessment of alternatives and ensures less susceptibility to local mistakes in individual comparisons. Although *Saaty* did not address group decision making in his seminal work on AHP [155], this topic can be found in his later works [158, 160]. Thus, following other MCDM techniques, AHP [89] can also be used for group decision making.

In their work, researchers dealing with group methods of decision making place great emphasis on the practical aspects of the techniques they develop, such as the decision context, examples, communication, cooperation or decision guidance [56, 63, 35, 134]. Sometimes the calculation aspect of the procedure is placed in the background.

In this chapter, we will not go too deeply into the practice of group decision making with AHP. Instead, we will try to look more at the preference aggregation methods that allow a final ranking to be created based on the opinions of various experts.

8.1   Aggregation methods

When considering group decision making with AHP, two basic issues need to be addressed [63]. The first one is the type of expert group and, as follows, the result aggregation strategy. The second point to consider is the mathematical procedure used for preference aggregation fitting of the given expert group and the results they provide.

In the literature, two kinds of expert groups are considered. The first one is referred to as a synergistic unit, which means a group of people who share the same values, have the same expertise and way of grading the significance of alternatives. The second type of group is a collection of individuals with perhaps different perceptions of the decision problem. In the first case, we can assume that people have a similar view regarding the significance of alternatives, and a similar understanding of verbal assessments such as "more important." In this situation, we can begin to aggregate individual comparisons. For example, if $C_p = [c_{ij}^{(p)}]$, $C_q = [c_{ij}^{(q)}]$ and $C_r = [c_{ij}^{(r)}]$ are the reciprocal PC matrices of the same size, we can create the matrix $C = [c_{ij}]$ where $c_{ij} = f(c_{ij}^{(p)}, c_{ij}^{(q)}, c_{ij}^{(r)})$ and $f : \mathbb{R}_+^3 \to \mathbb{R}_+$ is the aggregation function. In this approach, the ranking is calculated based on the PC matrix where the individual comparisons are the results of the aggregation of opinions from three experts p, q and r.

In the second case, we cannot aggregate individual judgments. It is possible that the experts have different understanding of the problem or different value systems. All this together can make the consideration of comparisons from different experts pointless. The solution is the aggregation of individual rankings (rather than comparisons). This is more like "numerical" voting, not agreeing on an opinion on the decision problem being considered. In the context of the considered example of three matrices C_p, C_q and C_r, the aggregation function $f : \mathbb{R}_+^3 \to \mathbb{R}_+$ aggregates the elements of priority vectors, i.e., $w(a_i) = f(w_p(a_i), w_q(a_i), w_r(a_i))$, where w is the aggregated ranking and w_p, w_q and w_r are the priority vectors coming from the p-th, q-th and r-th experts correspondingly.

8.1.1   Aggregation of individual judgments

The purpose of aggregating individual judgments (AIJ) is to propose for each pair of alternatives some "virtual", collective comparison that is a compromise between the assessments provided by different experts [63]. A natural candidate for this collective comparison is the mean of comparisons provided by all experts. Since the resulting matrix also needs to be a reciprocal PC matrix, ultimately, the only function suitable for aggregating individual comparisons is the geometric mean [2]. Thus, for $C_1, \ldots, C_q, \ldots, C_r$ PC matrices,

where

$$
C_q =
\begin{pmatrix}
1 & c_{1,2,q} & \cdots & \cdots & c_{1,n,q} \\
c_{2,1,q} & 1 & \cdots & \cdots & \vdots \\
\vdots & \cdots & \ddots & \cdots & \vdots \\
\vdots & \cdots & \cdots & \ddots & c_{n-1.n,q} \\
c_{n,1,q} & \cdots & \cdots & c_{n,n-1,q} & 1
\end{pmatrix},
$$

the aggregated matrix takes the form:

$$
C =
\begin{pmatrix}
1 & \left(\prod_{q=1}^{r} c_{1,2,q}\right)^{1/r} & \cdots & \left(\prod_{q=1}^{r} c_{1,n,q}\right)^{1/r} \\
\left(\prod_{q=1}^{r} c_{2,1,q}\right)^{1/r} & 1 & \cdots & \vdots \\
\vdots & \cdots & \ddots & \left(\prod_{q=1}^{r} c_{n-1.n,q}\right)^{1/r} \\
\left(\prod_{q=1}^{r} c_{n,1,q}\right)^{1/r} & \cdots & \cdots & 1
\end{pmatrix}.
$$

Then the matrix C is the subject of one of the priority calculation procedures described in Chapter 3. For example, using GMM, the priority of the i-th alternative for aggregated judgments is given as follows:

$$
w(a_i) = \left(\prod_{k=1}^{n}\left(\prod_{q=1}^{r} c_{i,k,q}\right)^{1/r}\right)^{1/n}. \tag{8.1}
$$

Example 12. *Let us consider the following four matrices:*

$$
C_1 =
\begin{pmatrix}
1 & 2 & \frac{4}{5} & \frac{9}{5} & 9 \\
\frac{1}{2} & 1 & \frac{1}{4} & \frac{7}{4} & 5 \\
\frac{5}{4} & 4 & 1 & \frac{7}{2} & 9 \\
\frac{5}{9} & \frac{4}{7} & \frac{2}{7} & 1 & 5 \\
\frac{1}{9} & \frac{1}{5} & \frac{1}{9} & \frac{1}{5} & 1
\end{pmatrix},
\quad
C_2 =
\begin{pmatrix}
1 & 2 & \frac{1}{4} & 1 & 9 \\
\frac{1}{2} & 1 & \frac{8}{9} & \frac{2}{3} & 8 \\
4 & \frac{9}{8} & 1 & 3 & 9 \\
1 & \frac{3}{2} & \frac{1}{3} & 1 & 9 \\
\frac{1}{9} & \frac{1}{8} & \frac{1}{9} & \frac{1}{9} & 1
\end{pmatrix},
$$

$$
C_3 =
\begin{pmatrix}
1 & 1 & \frac{3}{8} & \frac{5}{3} & 9 \\
1 & 1 & \frac{2}{5} & 1 & 9 \\
\frac{8}{3} & \frac{5}{2} & 1 & 6 & 9 \\
\frac{3}{5} & 1 & \frac{1}{6} & 1 & 5 \\
\frac{1}{9} & \frac{1}{9} & \frac{1}{9} & \frac{1}{5} & 1
\end{pmatrix},
\quad
C_4 =
\begin{pmatrix}
1 & 1 & \frac{3}{7} & \frac{7}{4} & 5 \\
1 & 1 & \frac{8}{9} & 1 & 7 \\
\frac{7}{3} & \frac{9}{8} & 1 & 2 & 9 \\
\frac{4}{7} & 1 & \frac{1}{2} & 1 & 5 \\
\frac{1}{5} & \frac{1}{7} & \frac{1}{9} & \frac{1}{5} & 1
\end{pmatrix}.
$$

provided by four experts assessing five alternatives: $a_1, \ldots, a_5$. The inconsistency of these matrices is $CI(C_1) = 0.028$, $CI(C_2) = 0.099$, $CI(C_3) = 0.048$

and $CI(C_4) = 0.023$ respectively. Thus, the aggregated matrix is given as

$$
C = \begin{pmatrix}
1 & \sqrt{2} & \frac{\sqrt{3}}{2^{3/4}\sqrt[4]{35}} & \frac{\sqrt[4]{21}}{\sqrt{2}} & 3\sqrt{3}\sqrt[4]{5} \\
\frac{1}{\sqrt{2}} & 1 & \frac{\sqrt[4]{2}}{2\sqrt[4]{\frac{2}{5}}} & \sqrt[4]{\frac{7}{6}} & 2^{3/4}\sqrt{3}\sqrt[4]{35} \\
\frac{2^{3/4}\sqrt[4]{35}}{\sqrt{3}} & \frac{3\sqrt[4]{\frac{5}{2}}}{2} & 1 & \sqrt{3}\sqrt[4]{14} & 9 \\
\frac{\sqrt{2}}{\sqrt[4]{21}} & \sqrt[4]{\frac{6}{7}} & \frac{1}{\sqrt{3}\sqrt[4]{14}} & 1 & \sqrt{35}3^{3/4} \\
\frac{1}{3\sqrt{3}\sqrt[4]{5}} & \frac{1}{2^{3/4}\sqrt{3}\sqrt[4]{35}} & \frac{1}{9} & \frac{1}{\sqrt{35}3^{3/4}} & 1
\end{pmatrix} =
$$

$$
= \begin{pmatrix}
1. & 1.414 & 0.423 & 1.513 & 7.77 \\
0.707 & 1. & 0.53 & 1.039 & 7.085 \\
2.361 & 1.886 & 1. & 3.35 & 9. \\
0.66 & 0.962 & 0.298 & 1. & 5.791 \\
0.1286 & 0.141 & 0.111 & 0.172 & 1.
\end{pmatrix}.
$$

Its consistency determined by Saaty's *CI is 0.019. The EVM-based ranking resulting from C is as follows:*

$$(0.222, 0.184, 0.408, 0.153, 0.031)^T. \tag{8.2}$$

The GMM ranking based on C is

$$(0.224, 0.185, 0.404, 0.154, 0.0308)^T. \tag{8.3}$$

8.1.2　Aggregation of individual priorities

The aggregation of individual priorities (AIP) approach delays the moment of judgment aggregation. In this case, therefore, the priority vectors are calculated first, separately for each expert's matrix, then these rankings are combined, producing a final assessment [63]. While for AIJ the geometric mean has to be used, for AIP either an arithmetic or geometric mean can be used as an aggregation function. This function does not have to be compatible with the reciprocity property, although it must meet the Pareto principle. According to this principle if for two alternatives a_i and a_j all r experts prefer a_i over a_j, i.e., for $c_{ijq} > 1$ and $q = 1, \ldots, r$, then by the group (after aggregation) a_i also needs to be preferred to a_j. It is easy to observe that both the arithmetic and geometric means meet this condition. Indeed, if $a_i \succeq_q a_j$, i.e., $w_q(a_i) \geq w_q(a_j)$ for $q = 1, \ldots, r$, where w_q is the priority vector of the q-th expert then both: the arithmetic mean

$$\frac{1}{n} \sum_{q=1}^{r} w_q(a_i) \geq \frac{1}{n} \sum_{q=1}^{r} w_q(a_j),$$

as well as the geometric mean

$$\left(\prod_{q=1}^{r} w_q(a_i)\right)^{\frac{1}{r}} \geq \left(\prod_{q=1}^{r} w_q(a_j)\right)^{\frac{1}{r}}$$

are compatible with the Pareto principle.

Example 13. *Let us consider the PC matrices $C_1, \ldots, C_4$ used in the Example 12. Their priority vectors computed using EVM are as follows:*

$$w_{ev}^{(1)} = \begin{pmatrix} 0.279 \\ 0.151 \\ 0.41 \\ 0.126 \\ 0.032 \end{pmatrix}, \quad w_{ev}^{(2)} = \begin{pmatrix} 0.199 \\ 0.179 \\ 0.407 \\ 0.188 \\ 0.025 \end{pmatrix},$$

$$w_{ev}^{(3)} = \begin{pmatrix} 0.195 \\ 0.181 \\ 0.47 \\ 0.123 \\ 0.028 \end{pmatrix}, \quad w_{ev}^{(4)} = \begin{pmatrix} 0.211 \\ 0.231 \\ 0.348 \\ 0.172 \\ 0.036 \end{pmatrix}.$$

The same individual rankings computed using GMM are:

$$w_{gm}^{(1)} = \begin{pmatrix} 0.283 \\ 0.15 \\ 0.407 \\ 0.126 \\ 0.032 \end{pmatrix}, \quad w_{gm}^{(2)} = \begin{pmatrix} 0.202 \\ 0.177 \\ 0.391 \\ 0.202 \\ 0.026 \end{pmatrix},$$

$$w_{gm}^{(3)} = \begin{pmatrix} 0.201 \\ 0.184 \\ 0.462 \\ 0.124 \\ 0.027 \end{pmatrix}, \quad w_{gm}^{(4)} = \begin{pmatrix} 0.209 \\ 0.232 \\ 0.348 \\ 0.172 \\ 0.037 \end{pmatrix}.$$

Thus, the aggregated ranking vectors are as follows:

$$w_{ev}^{+} = \begin{pmatrix} 0.221 \\ 0.186 \\ 0.409 \\ 0.153 \\ 0.03 \end{pmatrix}, \quad w_{ev}^{\times} = \begin{pmatrix} 0.219 \\ 0.184 \\ 0.407 \\ 0.15 \\ 0.03 \end{pmatrix},$$

$$w_{gm}^{+} = \begin{pmatrix} 0.224 \\ 0.186 \\ 0.402 \\ 0.156 \\ 0.031 \end{pmatrix}, \quad w_{gm}^{\times} = \begin{pmatrix} 0.221 \\ 0.184 \\ 0.4 \\ 0.153 \\ 0.031 \end{pmatrix},$$

where w_{ev}^{+}, $w_{ev}^{\times}$ are the aggregated $w_{ev}^{(1)}, \ldots, w_{ev}^{(4)}$ using the arithmetic and geometric mean respectively, and w_{gm}^{+}, $w_{gm}^{\times}$ are the aggregated $w_{gm}^{(1)}, \ldots, w_{gm}^{(4)}$ correspondingly. As the initial matrices were not very inconsistent and all four experts had a similar view as to the importance of the considered alternatives, the ranking vectors computed using different methods are quite similar. It is worth noting that since the input vectors $w_{ev}^{(1)}, \ldots, w_{ev}^{(4)}$ or $w_{gm}^{(1)}, \ldots, w_{gm}^{(4)}$ are scaled so that their entries sum up to one, then the aggregated vectors $w_{ev}^{+}, w_{ev}^{\times}, w_{gm}^{+}, w_{gm}^{\times}$ also are rescaled in such a way that all their components amount to one.

8.1.3   Opinions of varying importance

When aggregating assessments from different experts, we often assume that their opinions have the same importance (they have the same voting or decision power). Although such an assumption seems to be natural, it is not always true. Some experts may have more expertise and experience than others. Similarly, some policymakers may have greater power and competence than the rest. In such a case, the importance of the given expert must be reflected in the aggregation function. *Forman* and *Peniwati* [63] recommend using either a weighted arithmetic or weighted geometric mean. In the case of AIJ, the aggregated matrix takes the form

$$C = \begin{pmatrix} 1 & \left(\prod_{q=1}^{r} c_{1,2,q}^{\eta_q}\right)^{1/r} & \cdots & \left(\prod_{q=1}^{r} c_{1,n,q}^{\eta_q}\right)^{1/r} \\ \left(\prod_{q=1}^{r} c_{2,1,q}^{\eta_q}\right)^{1/r} & 1 & \cdots & \vdots \\ \vdots & \cdots & \ddots & \left(\prod_{q=1}^{r} c_{n-1,n,q}^{\eta_q}\right)^{1/r} \\ \left(\prod_{q=1}^{r} c_{n,1,q}^{\eta_q}\right)^{1/r} & \cdots & \cdots & 1 \end{pmatrix},$$

where $\eta_1 + \ldots + \eta_r = 1$, $\eta_1, \ldots, \eta_r \geq 0$ and η_q denotes the weight of the q-th expert. Based on the aggregated matrix the ranking is calculated. For example, the GMM priority vector's element (8.1) is as follows:

$$w(a_i) = \left(\prod_{k=1}^{n} \left(\prod_{q=1}^{r} c_{i,k,q}^{\eta_q}\right)^{1/r}\right)^{1/n}.$$

Similarly, when using AIP the ranking vectors are aggregated using the weighted arithmetic or geometric mean. Let

$$w^{(q)} = [w^{(q)}(a_1), \ldots, w^{(q)}(a_n)]^T$$

be the priority vector obtained from the PC matrix from the q-th expert. Thus, the aggregated priority vector

$$w = [w(a_1), \ldots, w(a_n)]^T$$

can be computed as

$$w(a_i) = \sum_{q=1}^{r} \eta_q \cdot w^{(q)}(a_i),$$

when the arithmetic mean is adopted as the aggregation function, or as

$$w(a_i) = \prod_{q=1}^{r} \left(w^{(q)}(a_i) \right)^{\eta_q},$$

when the geometric mean has been chosen by a facilitator. As before η_q means the "voting power" granted to the q-th expert.

8.2 Consensus methods

Intuitive AIP and AIJ are not the only methods that can be used for pairwise comparison-based group decision making. The calculation of a priority vector which is common for many experts can be perceived as a process of searching for a consensus. So, after defining what in the given case we consider as a consensus, we can turn the problem of aggregation into a problem of consensus seeking sometimes equivalent to the problem of optimization in relation to a certain consensus criterion. *Blagojevic* et al. [18] indicate that the first two methods are viable options when experts are willing to change their preferences during the decision-making process, and when outliers know considerably less than the other experts, while the last method performs well when experts want to stay with their initial opinions or outliers know more than the other members of the expert committee.

8.2.1 Consensus convergence model

An example of this way of thinking about the aggregation of preferences is the consensus convergence model (CCM) proposed by *Regan* et al. [150]. In CCM the priority of each alternative is calculated separately and the computational procedure is iterative.

In this approach, all the experts propose the initial priorities for the k-th alternative given as $w^{(0)} = \left(w_1^{(0)}(a_k), \ldots, w_r^{(0)}(a_k) \right)^T$ then the weights of respect matrix M is created

$$M = \begin{pmatrix} m_{11} & m_{12} & \cdots & m_{1r} \\ m_{21} & m_{22} & \cdots & m_{2r} \\ \vdots & \vdots & \ddots & \vdots \\ m_{r1} & m_{r2} & \cdots & m_{rr} \end{pmatrix},$$

where

$$m_{ij} = \frac{1 - \left| w_i^{(0)}(a_k) - w_j^{(0)}(a_k) \right|}{\sum_{j=1}^{n} 1 - \left| w_i^{(0)}(a_k) - w_j^{(0)}(a_k) \right|},$$

is the weight of respect that the i-th expert has for the j-th expert. The next set of priorities for the k-th alternative is computed as

$$w^{(s+1)} = Mw^{(s)}, \quad \text{for } s = 0, 1, \dots.$$

The procedure is repeated until a consensus is reached, i.e., the priorities from different experts are close enough to each other. In other words, iterations stop when $\left| w_i^{(l)}(a_k) - w_j^{(l)}(a_k) \right| < \epsilon$ for some small $\epsilon \in \mathbb{R}_+$ where $i, j = 1, \dots, r$ and $l \in \mathbb{N}_+$.

8.2.2 Geometric cardinal consensus index

Another way to reach a consensus has been proposed by Dong et al. [52]. In this approach, for every PC matrix $C_q = [c_{ij}^{(q)}]$ provided by the q-th expert the geometric cardinal consensus index (GCCI) is calculated.

$$GCCI(C_q) = \frac{2}{(n-1)(n-2)} \sum_{i<j} \left(\log \left(c_{ij}^{(q)} \right) - \log w(a_i) + \log w(a_j) \right),$$

where w is an aggregated priority vector computed using AIJ and the geometric mean method as the aggregation function. Then, iteratively, the matrices $C_1, \dots, C_r$ are modified so that $GCCI$ is small enough. The authors also provide acceptable thresholds of $GCCI$, which are $0.31, 0.35, 0.37$ for $n = 3$, $n = 4$ and $n > 4$ respectively [52, 18].

8.2.3 Group Euclidean distance minimization

The GED (group Euclidean distance) optimization model [18] is based on the idea of minimizing the GED function given as

$$GED = \left[\sum_{k=1}^{m} \sum_{i=1}^{n} \sum_{j=1}^{n} \left(c_{ij}^{(k)} - \frac{w(a_i)}{w(a_j)} \right)^2 \right]^{\frac{1}{2}},$$

where m is the number of experts, w is the aggregated priority vector and $c_{ij}^{(k)}$ is the result of comparisons from the k-th expert. According to this approach, the priority vector w minimizing GED is a reasonable compromise acceptable to all concerned. Formally, the optimization model is as follows:

$$\min f(w) = GED$$

$$s.t. \ \sum_{i=1}^{n} w(a_i) = 1 \text{ and}$$

$$w(a_i) > 0, \text{ for } i = 1, \ldots, n.$$

Many different numerical methods can be used to minimize GED. *Blagojevic* et al. [18] propose the simulated annealing method for this purpose, and call this approach SAAP (Simulated Annealing Aggregation Procedure). For instance, the results of *GED* minimization for $C_1, \ldots, C_4$ (Example 12), with the differential evolution optimization method [184] is as follows:

$$(0.254, 0.225, 0.313, 0.175, 0.032)^T.$$

Comparing the above result to GMM, EVM and AIJ (see 8.2, 8.3), it is clear that, although both rankings are ordinal identical, the cardinal differences are remarkable.

8.2.4 Other methods

In addition to the above three solutions, other consensus-based methods have also been proposed. *Escobar* and *Moreno-Jiménez* [58] proposed the aggregation of individual preference structures (AIPS) procedure. *Bryson* [35] developed a method using consensus-relevant information content. The GWLS model has been devised by *Sun* and *Greenberg* [186]. Another PD&R model comes from *Huang* et al. [83]. *Grošelj* et al. [75] consider the LP-GW-AHP model [81] as a method allowing priority aggregation from different experts. A good review of the recent efforts regarding the group decision-making methods using AHP and ANP is the work of *Ossadnik* et al. [138].

8.3 Compatibility index

When considering PC matrices from different experts, one may ask not only about their internal consistency but also about how they are mutually compatible. In a situation where there are several experts and each of them has a completely different opinion from the others it is difficult to make a decision. For this reason, in the literature on group decision making a lot of space is devoted to reaching a compromise. Very often, when there is a difference of opinion, experts are encouraged to discuss the problem and revise their opinions, if necessary, to achieve a joint view of the matter. To determine how far comparisons made by one expert differ from those of another, *Saaty* proposed a compatibility index [163]. His idea is based on the mutual comparison of corresponding elements in two PC matrices. Let $C_1 = [c_{ij}^{(1)}]$ and $C_2 = [c_{ij}^{(2)}]$ be two PC matrices from two different experts. Their opinion on the relationship between the i-th and j-th alternatives becomes more similar the more $c_{ij}^{(1)}$ and $c_{ij}^{(2)}$ are similar. In particular, if the assessments of both experts are identical then $c_{ij}^{(1)} = c_{ij}^{(2)}$, i.e., $c_{ij}^{(1)} \cdot c_{ji}^{(2)} = 1$. Based on this observation, the compatibility index was proposed as follows:

$$cmp(C_1, C_2) \stackrel{df}{=} \frac{1}{n^2} e^T C_1 \odot C_2^T e,$$

where e is the unit vector with n elements, i.e.,

$$e = \begin{pmatrix} 1 \\ \vdots \\ \vdots \\ 1 \end{pmatrix},$$

and $\odot$ is the Hadamard product, i.e.,

$$C_1 \odot C_2^T = \begin{pmatrix} 1 & c_{12}^{(1)} c_{21}^{(2)} & \cdots & c_{1n}^{(1)} c_{n1}^{(2)} \\ c_{21}^{(1)} c_{12}^{(2)} & 1 & \cdots & c_{2n}^{(1)} c_{n2}^{(2)} \\ \vdots & \vdots & \ddots & \vdots \\ c_{n1}^{(1)} c_{1n}^{(2)} & \cdots & c_{n,n-1}^{(1)} c_{n-1.n}^{(2)} & 1 \end{pmatrix}.$$

The compatibility index can also be written as

$$cmp(C_1, C_2) = \frac{1}{n^2} \sum_{i,j=1}^{n} c_{ij}^{(1)} c_{ji}^{(2)}.$$

It is easy to observe that for two identical PC matrices $C = C_1 = C_2$ it holds that $cmp(C, C) = 1$. In general, the smaller the compatibility index (the closer to 1) the more similar the two PC matrices are.

In [163] *Saaty* indicates an interesting relationship between the principal eigenvalue and compatibility index. It holds that

$$cmp(C, W) = \frac{\lambda_{max}}{n},$$

where $W = [w(a_i)/w(a_j)]$ and w is the priority vector computed using EVM. The above allows us to express the compatibility index using the consistency index, and reversely the consistency index using the compatibility index. These are:

$$\frac{n}{n-1}(cmp(C, W) - 1) = CI \text{ and } cmp(C, W) = \frac{n-1}{n} CI + 1.$$

Example 14. *Let us consider two matrices*

$$
C_1 = \begin{pmatrix}
1 & \boxed{\frac{1}{7}} & \boxed{6} & 7 & \frac{1}{5} \\
\boxed{7} & 1 & \frac{1}{6} & \frac{1}{5} & \frac{1}{3} \\
\boxed{\frac{1}{6}} & 6 & 1 & \frac{1}{8} & 6 \\
\frac{1}{7} & 5 & 8 & 1 & \frac{1}{7} \\
5 & 3 & \frac{1}{6} & 7 & 1
\end{pmatrix}, \quad
C_2 = \begin{pmatrix}
1 & \boxed{\frac{1}{3}} & \boxed{3} & 7 & \frac{1}{5} \\
\boxed{3} & 1 & \frac{1}{6} & \frac{1}{5} & \frac{1}{3} \\
\boxed{\frac{1}{3}} & 6 & 1 & \frac{1}{8} & 6 \\
\frac{1}{7} & 5 & 8 & 1 & \frac{1}{7} \\
5 & 3 & \frac{1}{6} & 7 & 1
\end{pmatrix},
$$

that differ in two comparisons. The first difference is a comparison between a_1 and a_2, while the second difference holds between a_1 and a_3. Their compatibility index is as follows:

$$
cmp(C_1, C_2) = \frac{1}{25}(1,1,1,1,1)
\begin{pmatrix}
1 & \frac{1}{3}\cdot 7 & 3\cdot\frac{1}{6} & 7 & 1 \\
3\cdot\frac{1}{7} & 1 & 1 & 1 & 1 \\
\frac{1}{3}\cdot 6 & 1 & 1 & 1 & 1 \\
1 & 1 & 1 & 1 & 1 \\
1 & 1 & 1 & 1 & 1
\end{pmatrix}
\begin{pmatrix}
1 \\ 1 \\ 1 \\ 1 \\ 1
\end{pmatrix},
$$

$$
cmp(C_1, C_2) = \frac{21 + \frac{1}{7}\cdot 3 + \frac{1}{3}\cdot 7 + \frac{1}{3}\cdot 6 + \frac{1}{6}\cdot 3}{25},
$$

hence

$$
cmp(C_1, C_2) = \frac{1103}{1050} = 1.05048.
$$

Chapter 9

Ordinal Inconsistency

AHP is a technique that uses quantitative pairwise comparisons. Alternatives are compared in a quantitative manner, i.e., the result of comparisons is a real number with a quantitative meaning. The numerical ranking that results from the priority-deriving method is also quantitative. Hence, if we had a pool of money to divide among the ranking participants, we could make such a division based on the numerical ranking, assuming that the priorities multiplied by 100% correspond to the percentage that each participant is expected to share. Very often, however, the results of pairwise comparison rankings are used differently. If we plan to buy a car and create a pairwise comparison ranking according to which we will make a purchase, we are only interested in the winner. Similarly, if we enter the competition for the best novel, the prize will go only to the best writer. In most cases, the podium has only three places and it does not matter whether the gold medalist overtook his predecessor significantly or by just a little. Such thinking reflects an ordinal (qualitative) approach. The ordinal perspective, quite popular recently [178, 28], is also present when the ranking is quantitative. Indeed, the quantitative pairwise comparison ranking is very often interpreted qualitatively. It is therefore justified to analyze quantitative pairwise comparison matrices in a qualitative way. This chapter will include some examples of how the qualitative approach can be used to analyze quantitative pair comparison matrices.

9.1   Condition of order preservation

9.1.1   Introduction

As mentioned before, the priority-deriving methods used in AHP assume that the proportions between the priorities of individual alternatives correspond to appropriate comparisons. Based on this assumption, we were able to prove Theorem 4 showing that when the PC matrix is consistent then $c_{ij} = w(a_i)/w(a_j)$. If the considered PC matrix is inconsistent, we can count at most on $c_{ij} \approx w(a_i)/w(a_j)$. From an ordinal perspective, however, the values that both sides of the equation take are important. If, for example, c_{ij} is greater than 1, i.e., the expert decided in a direct comparison that alternative a_i is more preferred than alternative a_j, then he will also expect that in the ranking the i-th alternative will come before the j-th alternative. Otherwise, he/she will probably question the results of the ranking. Based on this observation, *Bana e Costa* and *Vansnick* [11] formulated two, in essence ordinal, conditions (called COP – conditions of order preservation) related to the eigenvalue method. The authors' main complaint is that EVM does not meet these conditions, even when the inconsistency is "acceptably" low. Even worse, it is possible that, in parallel with the ranking violating COP, there may be a ranking that does not violate this condition. Although COP has been formulated with reference to EVM, both COP conditions apply to any quantitative PC priority-deriving method. Hence, the significance of this condition goes beyond the classic AHP and one priority calculation method.

9.1.2   Problem definition

COP is composed of two separate conditions [109]. The first one, called *the preservation of order preference condition (POP)*, claims that the nature of relationships between priorities in the priority vector should match individual comparisons. In other words, if the i-th alternative dominates the j-th alternative in the direct comparison, i.e., $c_{ij} > 1$, then it should also hold that:

$$w(a_1) > w(a_2). \tag{9.1}$$

The second condition, called *the preservation of order of intensity of preference condition (POIP)*, stipulates that if there are four alternatives such that in direct comparison a_i dominates a_j, more than a_p dominates a_q, i.e., $c_{ij} > c_{pq} > 1$, then this relationship transfers to the ranking, i.e.,

$$\frac{w(a_i)}{w(a_j)} > \frac{w(a_p)}{w(a_q)}. \tag{9.2}$$

The nature of both POP and POIP conditions is, in fact, qualitative. POP does not consider how much c_{ij} is greater than 1 and it does not require that in the same proportion $w(a_i)$ be greater than $w(a_j)$. Once again, it turns out that what counts for the expert is who the winner is and not the extent to which the winner overtakes the competitor. The second condition has quantitative elements. Preference intensities are compared. However, it is not important whether the difference between c_{ij} and c_{pq} is quantitatively comparable to the difference between the ratios $w(a_i)/w(a_j)$ and $w(a_p)/w(a_q)$. Thus, in essence, the second condition is also qualitative in nature.

Although both conditions of order preservation have been formulated for EVM, they remain valid for any quantitative priority-deriving method. Meeting *POP* and *POIP* seems to be natural for any priority vector.

Example 15. *To see COP in practice, let us consider the following PC matrix [11]:*

$$C = \begin{pmatrix} 1 & 2 & 3 & 5 & 9 \\ \frac{1}{2} & 1 & 2 & 4 & 9 \\ \frac{1}{3} & \frac{1}{2} & 1 & 2 & 8 \\ \frac{1}{5} & \frac{1}{4} & \frac{1}{2} & 1 & 7 \\ \frac{1}{9} & \frac{1}{9} & \frac{1}{8} & \frac{1}{7} & 1 \end{pmatrix}.$$

The priority vector calculated using EVM for C is

$$w = \begin{pmatrix} 0.426 \\ 0.281 \\ 0.165 \\ 0.101 \\ 0.0270 \end{pmatrix}.$$

Thus, it is easy to see that $c_{14} = 5$ and $c_{45} = 7$, thus, $c_{14} < c_{45}$ while

$$\frac{w(a_1)}{w(a_4)} = 4.231, \quad \frac{w(a_4)}{w(a_5)} = 3.75,$$

which means that

$$\frac{w(a_1)}{w(a_4)} > \frac{w(a_4)}{w(a_5)}.$$

Hence, the EVM violates the POIP condition for C. It is easy to verify that EVM is not the only method that violates the POIP condition in this case. GMM does not pass this test either. The Saaty consistency ratio for C is $CR(C) = 0.051$, so it is considered as acceptable.

Of course, it is possible that the POP condition also fails.

Example 16. *Let us consider the following PC matrix*

$$M = \begin{pmatrix} 1 & \frac{1}{7} & \frac{2}{3} & \frac{1}{9} & \frac{1}{7} \\ 7 & 1 & \frac{9}{2} & 1 & \frac{8}{9} \\ \frac{3}{2} & \frac{2}{9} & 1 & \frac{1}{3} & \frac{3}{7} \\ 9 & 1 & 3 & 1 & \frac{5}{4} \\ 7 & \frac{9}{8} & \frac{7}{3} & \frac{4}{5} & 1 \end{pmatrix}.$$

It is easy to verify that the EVM ranking is:

$$w = \begin{pmatrix} 0.042 \\ 0.299 \\ 0.085 \\ 0.309 \\ 0.264 \end{pmatrix}.$$

Hence, $c_{52} = 9/8 = 1.125$ while $w(a_5)/w(a_2) = 0.88222$, which is an evident violation of POP. GMM also fails in this case.

9.1.3 COP, error, and inconsistency

Since COP conditions seem to be an important ranking quality criterion, the question arises as to when a violation of the COP condition occurs and whether we can defend against it. The answer lies in the local inconsistency of the PC matrix. Let us look back to the maximum error criterion (Section 6.3.7) and prove the two following theorems [109]:

Theorem 10. *For every PC matrix $C = [c_{ij}]$ and the priority vector w, POP is met, i.e., for every $i, j = 1, \ldots, n$ it holds that*

$$c_{ij} > 1 \quad \text{implies} \quad w(c_i) > w(c_j) \tag{9.3}$$

if $\mathscr{E}_{ij} < \delta$ and $a_{ij} \geq \delta + 1$, for some $\delta > 0$, where $i, j = 1, \ldots, n$, and $i \neq j$.

Proof. Since $\mathscr{E}_{ij} < \delta$, thus by (Def. 28) every $e_{ji} - 1 < \delta$. Therefore, by the definition of e_{ij} (6.8) it holds that

$$\frac{1}{c_{ji}} \cdot \frac{w(a_j)}{w(a_i)} < \delta + 1, \tag{9.4}$$

hence

$$c_{ji} \frac{w(a_i)}{w(a_j)} > \frac{1}{\delta + 1}, \tag{9.5}$$

and due to the reciprocity

$$\frac{w(a_i)}{w(a_j)} > \frac{c_{ij}}{\delta + 1}. \tag{9.6}$$

Therefore, the ratio $w(a_i)/w(a_j)$ is strictly greater than one if only $c_{ij}/\delta+1 \geq 1$. Thus, if $\mathscr{E}_{ij} < \delta$ then $c_{ij} \geq \delta + 1$ implies $w(a_i) > w(a_j)$. As it is true for every $i, j = 1, \ldots n$ then $\delta > \mathscr{E}_{max}(C)$ and $c_{ij} \geq \delta + 1$ implies that POP is met. $\square$

The above Theorem clearly indicates that, to avoid violating POP, two conditions have to be met: the maximum error needs to be reasonably small (smaller than δ) and the experts in their opinions are strong enough (stronger than $\delta + 1$). In other words, if you do not want to be surprised by the result of the ranking, be consistent and determined in your judgment.

The situation is quite similar with the second POIP condition.

Theorem 11. *For the PC matrix C, and the priority vector w, POIP is met, i.e., for every $i, j = 1, \ldots, n$ it holds that*

$$c_{ij} > c_{kl} > 1 \quad implies \quad \frac{w(a_i)}{w(a_j)} > \frac{w(a_k)}{w(a_l)} \tag{9.7}$$

if $\mathscr{E}_{ij} < \delta$, $\mathscr{E}_{kl} < \delta$ and $c_{ij}/c_{kl} \geq (\delta + 1)^2$ for some $\delta > 0$, where $i, j, k, l = 1, \ldots .n$, and $i \neq j$, $k \neq l$.

Proof. Since $\mathscr{E}_{ij} < \delta$ and $\mathscr{E}_{kl} < \delta$, then by definition also $e_{ji} - 1 < \delta$ and $e_{kl} - 1 < \delta$. Thus, following the same reasoning as in Theorem 10 we get

$$\frac{w(a_i)}{w(a_j)} > \frac{c_{ij}}{\delta + 1} \quad and \quad \frac{w(a_l)}{w(a_k)} > \frac{c_{lk}}{\delta + 1},$$

i.e.,

$$\frac{w(a_i)}{w(a_j)} > \frac{c_{ij}}{\delta + 1} \quad and \quad \frac{w(a_k)}{w(a_l)} < c_{kl}(\delta + 1). \tag{9.8}$$

By dividing one inequality by the other, we get

$$\frac{\frac{w(a_i)}{w(a_j)}}{\frac{w(a_k)}{w(a_l)}} > \frac{\frac{c_{ij}}{\delta + 1}}{c_{kl}(\delta + 1)}. \tag{9.9}$$

This means that $w(a_i)/w(a_j)/w(a_k)/w(a_l)$ is greater than 1 if $c_{ij}/(\delta+1)/c_{kl}(\delta+1)$ is greater than or equal 1. As

$$\frac{\frac{c_{ij}}{\delta + 1}}{c_{kl}(\delta + 1)} = \frac{1}{(\delta + 1)^2} \cdot \frac{c_{ij}}{c_{kl}}$$

then, the truth of

$$\frac{c_{ij}}{c_{kl}} \geq (\delta + 1)^2 \tag{9.10}$$

implies that

$$\frac{w(a_i)}{w(a_j)} > \frac{w(a_k)}{w(a_l)}. \tag{9.11}$$

As the above holds for every $i, j, k, l = 1, \ldots n$ where $i \neq j$, $k \neq l$, $(i, j) \neq (k, l)$ and $\delta > 0$, which is equivalent to the fact that $\delta > \mathscr{E}_{max}(C)$ and $c_{ij}/c_{kl} \geq (\delta + 1)^2$, it implies that the POIP condition is met.

The meaning of the second theorem is similar to the first one. To meet POIP, we have to keep a low maximum error as well as a reasonably high preferential distance between alternatives. In fact, the higher the maximum error, the larger the difference in judgment between alternatives must be. In other words, we can afford even a substantial maximum error provided that the differences in priorities between the various options are sufficiently distinct. What is worth paying attention to is the locality of both the above theorems. It is possible that, for some (e.g., political) reasons, some alternatives are more sensitive than others. For example, the ranking of athletes, several of whom are multiple champions in a given discipline. In such a case, we may want to avoid violating POP and POIP with respect to these several champion candidates. Apart from these special options, we can tolerate some "small" violation of POP and POIP. If it is so, we just need to examine only these comparisons which are related to the "sensitive" alternatives. Keeping a low $\mathcal{E}_{ij}$ with respect to only a few alternatives is easier than taking into account all $n(n-1)/2$ local errors. Importantly, these recommendations do not depend on the priority-deriving method as the two above assertions do not make any assumptions as to the priority vector, except a fundamental one that the ratio $w(a_i)/w(a_j)$ should tend to the priority c_{ij}.

Theorems 10 and 11 link meeting the COP conditions with both the maximum error and the preferential distance between alternatives. The question arises, however, as to whether the fulfillment of the COP conditions cannot be made dependent on local inconsistency. The answer is positive. If the inconsistency index is "local" then this is indeed possible. It is worth noting here that the maximum error (as well as the local errors) is not the local index in the same sense as the Koczkodaj index is. In every single formula of $\mathcal{E}_{max}$ or $\mathcal{E}_{ij}$ the ranking result involved is not local by nature. The above suggests that for every inconsistency index there may be a formula using the preferential distance between alternatives that allows the expert (or facilitator) to decide whether the COP condition will be violated or not. Let us look at the local inconsistency index $K(C)$ in the light of the COP criteria. $\square$

Theorem 10 prompts us to make another assertion similar to those provided in [110]. The new Theorem, however, needs an assumption that the priority-deriving method is either EVM or GMM.

Theorem 12. *For the EVM- or GMM-based ranking vector w, the PC matrix $C = [c_{ij}]$, it holds that*

$$c_{ij} > \frac{1}{1 - K(C)} \quad implies \quad w(a_i) > w(a_j). \tag{9.12}$$

Proof. From 10 we obtain

$$\{(\mathcal{E}_{max}(C) \le \delta) \Rightarrow (c_{ij} > \delta + 1)\} \Rightarrow \{(c_{ij} > 1) \Rightarrow (w(a_i) > w(a_j))\} \tag{9.13}$$

for some $\delta \in \mathbb{R}_+$. Due to the (6.47) the right side of (9.13) is

$$\left(\mathcal{E}_{max}(C) \le \frac{1}{1 - K(C)} - 1\right) \Rightarrow \left(c_{ij} > \frac{1}{1 - K(C)}\right) \tag{9.14}$$

and since $\frac{1}{1-K(C)} > 1$ then also

$$c_{ij} > \frac{1}{1 - K(C)} \quad implies \quad w(a_i) > w(a_j) \tag{9.15}$$

which is the desired assertion. If the right side of (9.12) holds for every $i, j = 1, \ldots, n$ then POP is met. $\qquad \square$

Similarly, the second *POIP* theorem leads to a new interesting assertion [110].

Theorem 13. *For the EVM- or GMM-based ranking vector w, the PC matrix $C = [c_{ij}]$, it holds that:*

$$\frac{c_{ij}}{c_{kl}} > \left(\frac{1}{1 - K(C)}\right)^2 \quad implies \quad \frac{w(a_i)}{w(a_j)} > \frac{w(a_k)}{w(a_l)}. \tag{9.16}$$

Proof. From 11 it holds that

$$\left\{ (\mathscr{E}_{max}(C) \le \delta) \Rightarrow \frac{c_{ij}}{c_{kl}} > (\delta + 1)^2 \right\} \Rightarrow \left\{ c_{ij} > c_{kl} > 1 \Rightarrow \frac{w(a_i)}{w(a_j)} > \frac{w(a_k)}{w(a_l)} \right\} \tag{9.17}$$

for some $\delta \in \mathbb{R}_+$. Due to the (6.47) the right side of (9.17) is:

$$\left(\mathscr{E}_{max}(C) \le \frac{1}{1 - K(C)} - 1 \right) \Rightarrow \frac{c_{ij}}{c_{kl}} > \left(\frac{1}{1 - K(C)} - 1 + 1 \right)^2 \tag{9.18}$$

Since $\frac{1}{1-K(C)} > 1$, then also

$$\frac{c_{ij}}{c_{kl}} > \left(\frac{1}{1 - K(C)} \right)^2 \Rightarrow \frac{w(a_i)}{w(a_j)} > \frac{w(a_k)}{w(a_l)} \tag{9.19}$$

which is the desired assertion. If the right side of (9.16) holds for every $i, j, k, l = 1, \ldots, n$ then POIP is met. $\qquad \square$

The two theorems above are, in a sense, a reflection of the previous assertions, 10 and 11. Here there is also a kind of trade-off between inconsistency and the preferential distance between alternatives. In this case, however, the inconsistency index is completely local. This clearly suggests that meeting the COP conditions can be highly dependent of the local inconsistency and the preferential distance between alternatives. Attempting to prove (or disprove) the above theorems for non-local inconsistency indices (e.g., *CI*) could be an interesting and useful exercise.

9.2   Kendall–Babington Smith inconsistency index

The pairwise comparison method goes far beyond the AHP introduced by *Saaty* in his seminal paper *"A scaling method for priorities in hierarchical structures"* [155]. Some attribute the first written sources treating pairwise comparison as a decision-making method to *Ramon Llull* [41]. Indeed *Llull*, coming from Palma de Majorca, an alchemist, mathematician, theologian and monk, proposed in the late thirteenth century an election system based on the qualitative comparing of alternatives. Several centuries later, *Condorcet* re-invented Llull's solution [42]. Once again, Llull's method was reinvented in the twentieth century by *Copeland* [43, 154]. In all these cases, the nature of comparisons is qualitative. Experts (voters) can indicate the better of the two but they cannot say how much more they prefer the winner. Of course, in addition to these three, *Llull*, *Condorcet* and *Copeland*, there are a number of works on social choice and welfare theory dealing with qualitative pairwise comparisons [187].

In addition, in the case of a qualitative approach, one can consider the consistency of a set of pairwise comparisons. The classic way to deal with this task is the method proposed by *Kendall* and *Babington Smith* based on the number of inconsistent triads in relation to their maximum number. The only problem with this approach is that it does not take ties into account. For this reason, the original coefficient could not be applied to AHP, as 1 is the result of a comparison representing a tie. In this section, we look at the idea behind the original Kendall–Babington Smith coefficient and its extension.

The problem of incompatibility between the Kendall–Babington Smith index and AHP was pointed out by *Jensen* and *Hicks* [92]. They also proposed to extend the Kendall–Babington Smith index to rankings with ties. The distribution of the number of circular (inconsistent) triads was studied by *Iida* [86] and *Alway* [5]. *Iida* also proposed an ordinality consistency test [85]. The relationships between cardinal and ordinal rankings were the subject of research conducted by *Genest* et al. [71], *Jensen* [91], *Kéri* [96] and *Brunelli* [28]. *Kéri* also classified different types of triads. A graph theoretical approach to the analysis of PC matrices is represented by *Gass* [68].

9.2.1   Ordinal inconsistency

In order to define ordinal consistency, let us introduce several formal notions related to graph theory which help us to represent ordinal pairwise comparisons.

Definition 32. *A family of* tournaments *with n vertices will be denoted by $\mathscr{T}_n^t$, where $\mathscr{T}_n^t = \{(V, E_d)$ is a t-graph, where $|V| = n\}$, and similarly, a family of* generalized tournaments *with n vertices will be denoted by $\mathscr{T}_n^g$, where $\mathscr{T}_n^g = \{(V, E_u, E_d)$ is a gt-graph, where $|V| = n\}$.*

It is easy to verify that for every $n > 1$ it holds that $\mathcal{T}_n^t \subsetneq \mathcal{T}_n^g$.

Definition 33. *A family of generalized tournaments with n vertices and m directed edges will be denoted as $\mathcal{T}_{n,m}^g = \{(V, E_u, E_d)$ which is a gt-graph, where $|V| = n$ and $|E_d| = m\}$.*

The above two definitions will allow us to shorten some theorems and statements. For example, the statement 'X is a generalized tournament with 5 vertices and 7 directed edges' is equivalent to $X \in \mathcal{T}_{5,7}^g$, etc.

Definition 34. *The individual vertex a is said to be contained by the triad $t = \{a_i, a_k, a_j\}$ if $a \in t$. The triad $t = \{a_i, a_k, a_j\}$ is said to be covered by the edge $(p,q) \in E_d$ if $p, q \in t$.*

Definition 35. *A double tournament (also called a dt-graph), is a generalized tournament $G = (V_1 \cup V_2, E_{d_1} \cup E_{d_2}, E_u)$ such that (V_1, E_{d_1}) and (V_2, E_{d_2}) are tournaments, where $V_1 \cap V_2 = \emptyset$ and $E_u = \{\{c, d\} : c \in V_1, d \in V_2\}$.*

Although we usually denote directed and undirected edges in between two vertices a_i and a_j as (a_i, a_j) and $\{a_i, a_j\}$ correspondingly, due to the intuitive association with the plot of a graph we can also denote them as $a_i \to a_j$ or $a_i - a_j$. Depending on the context, we will use either of the notations.

Kendall and Babington Smith considered ordinal pairwise comparisons without ties [95]. In other words, their ordinal PC matrices (Sec. 2.4) have no zeros. To define inconsistency, they adopt the transitivity of the preference relation considered with respect to triads. They noticed that it is natural to expect that if the i-th alternative is more preferred than the k-th alternative and the k-th alternative is more preferred than the j-th alternative, then the i-th alternative should be more preferred during the direct comparison than the j-th alternative. This kind of triad is considered consistent, otherwise the triad is inconsistent (contradictory). It is easy to observe that for three alternatives without ties there are only two possible types of triads: the consistent triad $a_i \succ a_k, a_k \succ a_j$ and $a_i \succ a_j$ hereinafter abbreviated to as CT_3, and the inconsistent (contradictory, circulant) triad $a_i \succ a_k, a_k \succ a_j$ and $a_j \succ a_i$.

The greater the number of inconsistent triads, the less reliable the ranking. Therefore, to determine the degree of inconsistency, *Kendall and Babington Smith* [95] calculate the maximal number of inconsistent triads (Fig. 9.1) in an n-vertex tournament from $\mathcal{T}_n^t$. Let T_C be the *t-graph* of C and let the actual number of inconsistent triads in T_C be given by $|T_C|_i$, and the maximal possible number of inconsistent triads for the n-vertex tournament as $\mathcal{I}(n)$, where

$$\mathcal{I}(n) \stackrel{df}{=} \begin{cases} \frac{n^3 - n}{24} & \text{when } n \text{ is odd} \\ \frac{n^3 - 4n}{24} & \text{when } n \text{ is even} \end{cases}. \tag{9.20}$$

Then, the Kendall–Babington Smith index for C defined in [95] is given:

$$\zeta(C) \stackrel{df}{=} 1 - \frac{|T_C|_i}{\mathcal{I}(n)}. \tag{9.21}$$

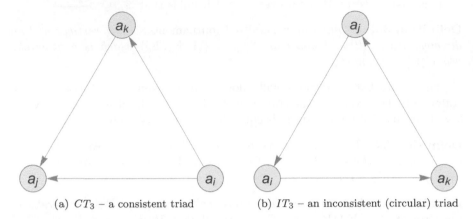

(a) CT_3 – a consistent triad   (b) IT_3 – an inconsistent (circular) triad

Figure 9.1: Triads for ordinal pairwise comparisons without ties

The definition of the index is accompanied by a description of how such a maximally inconsistent tournament may look. It turns out that it is a graph where the in-degrees of vertices are evenly distributed. Below we can see two such maximally inconsistent tournaments (Fig. 9.2).

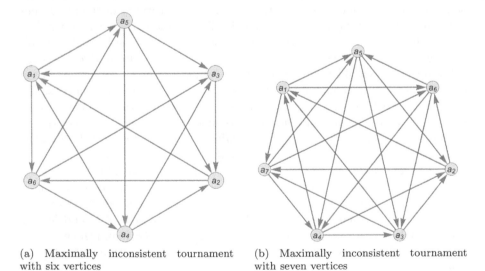

(a) Maximally inconsistent tournament with six vertices   (b) Maximally inconsistent tournament with seven vertices

Figure 9.2: An example of the most inconsistent tournament with six and seven vertices

When we allow for ties, the number of possible types of triads increases. Besides CT_3 and IT_3 there are another five we have to deal with:

CT_0 – consistent triad, all three alternatives are equally preferred, i.e., $a_i \sim a_k, a_k \sim a_j$ and $a_i \sim a_j$.

IT_1 – inconsistent triad such that $a_i \sim a_k, a_k \sim a_j$ and $a_i \prec a_j$.

IT_2 – inconsistent triad such that $a_i \sim a_k, a_k \prec a_j$ and $a_j \prec a_i$.

CT_{2a} – consistent triad such that $a_i \sim a_k, a_k \prec a_j$ and $a_i \prec a_j$.

CT_{2b} – consistent triad such that $a_i \sim a_k, a_j \prec a_k$ and $a_j \prec a_i$.

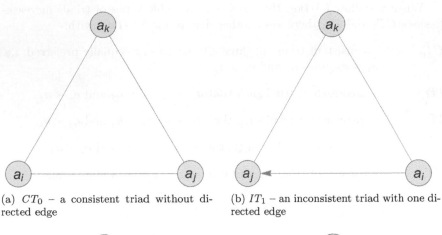

(a) CT_0 – a consistent triad without directed edge

(b) IT_1 – an inconsistent triad with one directed edge

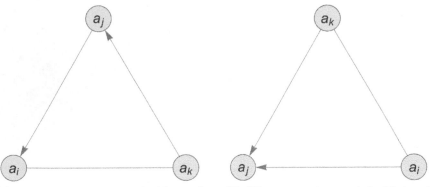

(c) IT_2 – an inconsistent triad with two directed edges

(d) CT_{2a} – a consistent triad with two directed edges variant (a)

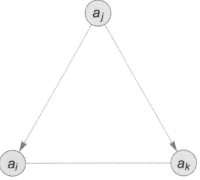

(e) CT_{2b} – a consistent triad with two directed edges variant (b)

Figure 9.3: Extra triads introduced by adding ties

The above triads can be easily represented as tournaments with ties (Figure 9.3). The increased number of types of triads makes the maximal number of inconsistent triads higher. For example, $\mathcal{I}(4) = 2$. It is easy to verify that, in fact, this number must be higher. Let us look at the graph in (Fig 9.4).

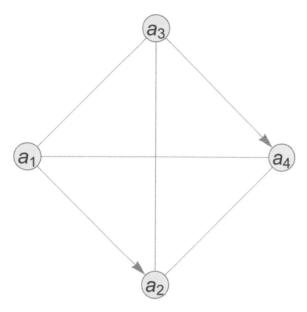

Figure 9.4: Generalized tournament with 4 vertices and four IT_1 triads

It contains four inconsistent triads of the type IT_1. They are $\{c_1 \to c_2, c_2 - c_3, c_3 - c_1\}$, $\{c_1 \to c_2, c_2 - c_4, c_4 - c_1\}$, $\{c_1 - c_3, c_3 \to c_4, c_4 - c_1\}$, and $\{c_2 - c_3, c_3 \to c_4, c_4 - c_1\}$. Thus, it is clear that the formulas (9.20) and (9.21) do not cover ordinal pairwise comparisons with ties.

In order to extend the index $\zeta(C)$ for ordinal matrices with ties, *Kułakowski* [113] provided a formula describing the maximal number of inconsistent triads in a generalized tournament with n vertices:

$$\mathcal{Y}(n) = \binom{n}{3} - \left(\binom{\lfloor \frac{n}{2} \rfloor}{3} - \mathcal{I}\left(\left\lfloor \frac{n}{2} \right\rfloor\right) \right) - \left(\binom{\lceil \frac{n}{2} \rceil}{3} - \mathcal{I}\left(\left\lceil \frac{n}{2} \right\rceil\right) \right). \quad (9.22)$$

The above allowed the generalized Kendall–Babington Smith index for C [113] to be formulated as

$$\zeta_g(C) \overset{df}{=} 1 - \frac{|G_C|_i}{\mathcal{Y}(n)}. \quad (9.23)$$

where $|G_C|_i$ is the number of inconsistent triads in the graph of the ordinal PC matrix with ties (Def. 13). The formula (9.22) can be simplified to:

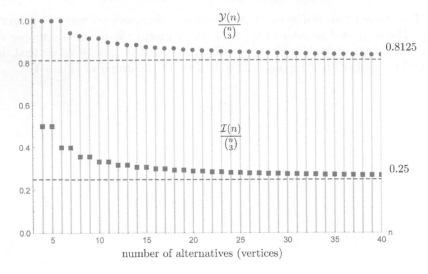

Figure 9.5: Maximum levels of inconsistent triads in tournaments and generalized tournaments

$$\mathcal{Y}(n) = \begin{cases} \frac{13n^3 - 24n^2 - 16n}{96} & \text{when } n \bmod 4 = 0 \text{ and } n \geq 4 \\ \frac{13n^3 - 24n^2 - 19n + 30}{96} & \text{when } n \bmod 4 = 1 \text{ and } n \geq 5 \\ \frac{13n^3 - 24n^2 - 4n}{96} & \text{when } n \bmod 4 = 2 \text{ and } n \geq 6 \\ \frac{13n^3 - 24n^2 - 19n + 18}{96} & \text{when } n \bmod 4 = 3 \text{ for } n \geq 3 \end{cases}.$$

It is easy to observe that in most cases (except the smallest graphs) the maximal number of inconsistent triads in a tournament and a generalized tournament is significantly smaller that the number of all possible triads. This means that we are usually not able to construct a "totally ordinal inconsistent" PC matrix or graph. However, when ties are allowed the number of inconsistent triads that may appear in the graph increases. Indeed, when we compare the ratios $\mathcal{I}(n)/\binom{n}{3}$ and $\mathcal{Y}(n)/\binom{n}{3}$ the latter is significantly larger (Fig. 9.5).

Interestingly, as the percentage of the maximum number of inconsistent triads decreases as the graph size increases, it never falls below a certain level. For tournaments, this level is

$$\lim_{n \to \infty} \frac{\mathcal{I}(n)}{\binom{n}{3}} = 0.25,$$

while for generalized tournaments, it is

$$\lim_{n \to \infty} \frac{\mathcal{Y}(n)}{\binom{n}{3}} = 0.8125.$$

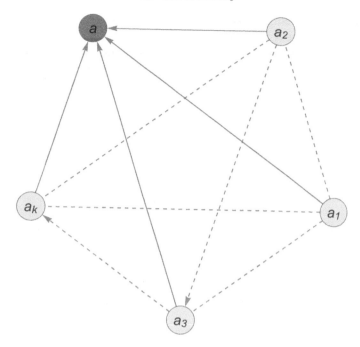

Figure 9.6: Consistent triads introduced by $a \in V$ with $\deg_{in}(a) = k$

9.2.2 Number of inconsistent triads without ties

Although we know the maximal number of triads in the case when ties are not allowed, here we will follow how the formula (9.20) is constructed. The experience gathered will allow us to later expand this formula so that it also includes inconsistent triads with ties.

Theorem 14. *Let $G = (V, E_u, E_d) \in \mathscr{T}_n^g$. Then every vertex $a \in V$, for which $\deg_{in}(a) = k$, is contained by at least $\binom{k}{2}$ consistent triads of the type CT_{2a} or CT_3. Those triads are said to be introduced by a.*

Proof. Let $a_1, \ldots, a_k \in V$ be the vertices such that the edges $a_i \to a$ are in E_d. Since G is a *gt-graph* with n vertices, then for every a_i, a_j for $i, j = 1, \ldots, k$ there must exist an edge $a_i \to a_j$, $a_j \to a_i$ in E_d or $a_i - a_j$ in E_u. In the first two cases, a_i, a, a_j make a consistent triad type CT_{2a}, while in the third case a_i, a, a_j form a consistent triad type CT_3. Since there are k vertices adjacent to the incoming edge to a there are at least as many different consistent triads containing a as doubletons composed of $a_1, \ldots, a_k$, i.e., $\binom{k}{2}$. See (Fig. 9.6). $\square$

In fact, a can form more consistent triads than suggested by the theorem above. This is because there may be more edges in the form $a \to a_{k+1}, \ldots, a \to a_{k+r}$. Thus, in G there may also be some triads of the type CT_{2b} containing a.

The above theorem is also true for tournaments. However, as the only consistent triad in the tournament is CT_3, the only consistent triads containing a are those introduced by a. From this, we understand the following corollary.

Corollary 2. *For $T = (V, E_d) \in \mathscr{T}_n^t$ every vertex $a \in V$ with $deg_{in}(a) = k$ is contained by $\binom{k}{2}$ triads of the type CT_3.*

In other words, in order to construct a tournament with the maximal number of inconsistent triads, one needs to minimize the number of consistent triads introduced by elements of V, i.e.,

$$|T|_c \stackrel{df}{=} \sum_{a \in V} \binom{deg_{in}(a)}{2}. \tag{9.24}$$

As the total number of triads is $\binom{n}{3}$, then the number of inconsistent triads is given as

$$|T|_i = \binom{n}{3} - |T|_c,$$

i.e.,

$$|T|_i = \binom{n}{3} - \sum_{c \in V} \binom{deg_{in}(a)}{2}. \tag{9.25}$$

As the undirected graph $P = (V, E)$ holds $\sum_{a \in V} deg(a) = 2|E|$ [49, p. 5], then

$$\sum_{a \in V} deg_{in}(a) = |E| = \binom{n}{2}. \tag{9.26}$$

Thus, our goal is to maximize $|T|_i$ (minimize $|T|_c$). Before we deal with it, we need two auxiliary definitions. Their role is similar to the definitions 32 and 33. They will help us shorten some of our reasoning.

Definition 36. *A gt-graph is said to be maximal if it has the maximal possible number of inconsistent triads among the gt-graphs with the same number of vertices. Families of the maximal tournaments and general tournaments will be denoted as follows:*

$$\overline{\mathscr{T}_n^t} = \{T \in \mathscr{T}_n^t \text{ such that } |T|_i = \max_{T_r \in \mathscr{T}_n^t} |T_r|_i\}, \tag{9.27}$$

$$\overline{\mathscr{T}_n^g} = \{G \in \mathscr{T}_n^g \text{ such that } |G|_i = \max_{G_r \in \mathscr{T}_n^g} |G_r|_i\}. \tag{9.28}$$

Before we prove the theorem about the construction of the maximal tournament (and as follows a maximally inconsistent set of ordinal comparisons without ties), let us look at two diophantine identities. For $r \in \mathbb{N}_+$ it holds that:

$$\binom{2r+1}{2} = r \cdot (2r+1) \tag{9.29}$$

and

$$\binom{2r}{2} = r \cdot r + r(r-1). \tag{9.30}$$

The first equality (9.29) means that for a graph with $n = 2r+1$ vertices, it is possible to assign exactly r incoming edges to every vertex a in V when n is odd. Similarly (9.30), providing that $n = 2r$ is even, it is possible to assign r incoming edges to r vertices and $r-1$ incoming edges to the other r vertices.

Theorem 15. *The number of inconsistent triads in the tournament $T = (V, E_d)$ is maximal, i.e., $T \in \overline{\mathscr{T}_n^t}$, if and only if*

1. *for every $a \in V$ $deg_{in}(a) = r$ when the number of vertices is odd and equals $n = 2r+1$, otherwise*

2. *there are r vertices $a_1, \ldots, a_r$ in V such that $deg_{in}(a_i) = r$, and r vertices $a_{r+1}, \ldots, a_n$ such that $deg_{in}(a_j) = r-1$, where $n = 2r$ and $1 \leq i \leq r < j \leq n$.*

Proof. To prove the theorem, we need to show that (9.24) is minimal when the arrangement of degrees of vertices corresponds to the one mentioned in the thesis of the theorem. Let us consider the n odd case first. Suppose that $n = 2r+1$ and (9.24) is minimal, but not all the vertices have input degrees equal r. Hence, there must be at least one $a_i \in V$ such that $deg_{in}(a_i) \neq r$. Let us suppose that $p = deg_{in}(a_i) > r$ (the case when $p < r$ is symmetric). The expressions (9.26) and (9.29) imply that there must exist at least one $a_j \in V$ such that $deg_{in}(a_j) = q < r$. Therefore, we can decrease p and increase q by one without changing the sum (9.26) just by changing the direction of the edge $a_j \to a_i$ to $a_i \to a_j$. Since $p + q = z$ and z is constant, the sum of consistent triads introduced by a_i and a_j is given as:

$$\binom{p}{2} + \binom{q}{2} = \binom{p}{2} + \binom{z-p}{2} = p(p-z) + \frac{z(z-1)}{2}. \tag{9.31}$$

Since $z(z-1)/2$ is constant, let

$$f(p) \overset{df}{=} p(p-z) + \frac{z(z-1)}{2}. \tag{9.32}$$

A decrease in p entails a decrease in $f(p)$ if

$$f(p) - f(p-1) > 0 \tag{9.33}$$

which holds if and only if

$$2p > (z-1). \tag{9.34}$$

As $p > q$ and $p + q = z$ then also (9.34) holds, hence, by decreasing $\deg_{in}(c_i)$ and increasing $\deg_{in}(c_j)$ by one, we can decrease the value of (9.24). This fact is contrary to the assumption that (9.24) is minimal. In conclusion, we get that all vertices should have an input degree equal to r.

The proof for $n = 2r$ is similar to the case when $n = 2r + 1$. The only difference is that as a_i we should adopt such a vertex for which $\deg_{in}(a_i) \neq r$ and $\deg_{in}(a_i) \neq r - 1$. Note that there must be at least one if we (by contradiction) reject the thesis of the theorem, and simultaneously allege that (9.24) is minimal. $\qquad\square$

The proof of (Theorem 15) also suggests an algorithm allowing the conversion of any tournament into the maximal tournament. In every step of this algorithm, it would be enough to find a_i whose in-degree differs from r (when n is odd) or differs from r and $r - 1$ (when n is even) and decreases (or increases) its input degree together with increasing (or decreasing) the input degree of its counterpart a_j. The algorithm would stop when there is no such pair (a_i, a_j). The algorithm stops as with every iteration the number of inconsistent triads in a graph increases but obviously it cannot be greater than $\binom{n}{3}$.

Kendall and *Babington Smith* [95] suggested the construction of the most inconsistent graphs resembling the creation of *circulant graphs* [141]. Namely, initially a graph consists of the cycle $a_1 \rightarrow a_2 \rightarrow a_3 \rightarrow \ldots \rightarrow a_n \rightarrow a_1$, then another cycle is added $a_1 \rightarrow a_3 \rightarrow a_5 \rightarrow \ldots \rightarrow a_n \rightarrow a_2 \rightarrow \ldots$ if n is even, or two cycles $a_1 \rightarrow a_3 \rightarrow \ldots \rightarrow a_{n-1} \rightarrow a_1$ and $a_2 \rightarrow a_4 \rightarrow \ldots \rightarrow a_n \rightarrow a_2$ if n is odd, and so on. Adding cycles needs to be continued until adding all $\binom{n}{2}$ edges. Examples of the maximally inconsistent tournaments $T_X \in \mathscr{T}_6^t$ and $T_Y \in \mathscr{T}_7^t$ are shown in (Fig. 9.2). The matrices X and Y corresponding to these graphs are as follows:

$$
X = \begin{pmatrix}
0 & 1 & 1 & 1 & -1 & -1 \\
-1 & 0 & 1 & 1 & 1 & -1 \\
-1 & -1 & 0 & 1 & 1 & 1 \\
-1 & -1 & -1 & 0 & 1 & 1 \\
1 & -1 & -1 & -1 & 0 & 1 \\
1 & 1 & -1 & -1 & -1 & 0
\end{pmatrix},
$$

$$
Y = \begin{pmatrix}
0 & 1 & 1 & 1 & -1 & -1 & -1 \\
-1 & 0 & 1 & 1 & 1 & -1 & -1 \\
-1 & -1 & 0 & 1 & 1 & 1 & -1 \\
-1 & -1 & -1 & 0 & 1 & 1 & 1 \\
1 & -1 & -1 & -1 & 0 & 1 & 1 \\
1 & 1 & -1 & -1 & -1 & 0 & 1 \\
1 & 1 & 1 & -1 & -1 & -1 & 0
\end{pmatrix}.
$$

The construction Theorem 15 indicates how to create maximum graphs. However, the question remains as to how such graphs relate to the formula proposed by Kendal and Babington Smith (9.20). Let us try to calculate it using equation (9.25) as the starting point.

Theorem 16. *For every maximal tournament* $T = (V, E_d)$, $n \geq 3$ *with the form defined by the Theorem 15, it holds that*

$$|T|_i = \mathcal{I}(n). \tag{9.35}$$

Proof. Let $n = 2r + 1$ and $r \in \mathbb{N}_+$. According to (9.25)

$$|T|_i = \binom{2r + 1}{3} - \sum_{c \in V} \binom{\deg_{in}(c)}{2}. \tag{9.36}$$

Then, due to (Theorem 15)

$$|T|_i = \binom{2r + 1}{3} - \underbrace{\left(\binom{r}{2} + \ldots + \binom{r}{2} \right)}_{2r+1}, \tag{9.37}$$

$$|T|_i = \frac{r(2r - 1)(2r + 1)}{3} - \frac{(r - 1)r(2r + 1)}{2}, \tag{9.38}$$

$$|T|_i = \frac{r\left(2r^2 + 3r + 1\right)}{6} = \frac{(2r + 1)^3 - (2r + 1)}{24}, \tag{9.39}$$

$$|T|_i = \frac{(2r + 1)^3 - (2r + 1)}{24} = \frac{n^3 - n}{24} = \mathcal{I}(n). \tag{9.40}$$

Similarly, when $n = 2r$ and $r \in \mathbb{N}_+$. Then, due to (Th. 15)

$$|T|_i = \binom{2r}{3} - \underbrace{\left(\binom{r}{2} + \ldots + \binom{r}{2} \right)}_{r} - \underbrace{\left(\binom{r - 1}{2} + \ldots + \binom{r - 1}{2} \right)}_{r}, \tag{9.41}$$

$$|T|_i = \frac{r(2r - 2)(2r - 1)}{3} - \frac{(r - 1)r^2}{2} - \frac{(r - 2)(r - 1)r}{2}, \tag{9.42}$$

$$|T|_i = \frac{r\left(r^2 - 1\right)}{3} = \frac{(2r)^3 - 4(2r)}{24} = \frac{n^3 - 4n}{24} = \mathcal{I}(n), \tag{9.43}$$

which satisfies the thesis of the theorem. $\qquad\square$

9.2.3   Number of inconsistent triads with ties

This section is divided into three subsections. In the first of them, we will laboriously construct the function $\mathcal{H}(n, m)$, which will be the upper limit of the number of inconsistent triads in any generalized tournament $G \in \mathcal{T}_{n,m}^{g}$ with n vertices and m directed edges. This function will allow us to assess the extent to which the $n \times n$ PC matrix containing $n - m$ ties can be inconsistent (in terms of the number of inconsistent triads). In the second subsection, we construct the generalized tournament with n vertices which will be as inconsistent as the maximal value of $\mathcal{H}(n, m)$ for a fixed n and $m = 0, 1, \ldots, \binom{n}{2}$. Then, we simply conclude that this graph is the maximally inconsistent generalized tournament with n vertices. We also provide the formulas allowing computation of the maximal number of inconsistent triads. In the last subsection, we summarize the results and show that when starting with a certain size of graphs (sets of ordinal pairwise comparisons) they can never be completely inconsistent.

9.2.3.1   Limit the number of inconsistent triads

Probably the simplest example of a consistent ordinal generalized tournament is a fully undirected graph with n vertices. Indeed, it contains only consistent triads. Thus, by adding or replacing undirected edges with directed ones, we are making a graph increasingly inconsistent. Of course, when we replace all the undirected edges by their directed counterparts, we convert a generalized tournament into an ordinary tournament, whose maximal number of inconsistent triads is given by $\mathcal{I}(n)$. However, we know from experiments that the number of inconsistent triads in a generalized tournament of a given size is higher than in an ordinary tournament (see Fig. 9.4). Hence, replacing all undirected edges with directed ones is probably not the best strategy. These two intuitive observations lead to the conclusion that the maximal generalized tournament should not contain too many consistent triads of the type CT_{2a} and CT_{2b} and not too few directed edges in order to eliminate consistent triads of the type CT_0. This leads to the following theorem.

Theorem 17. *Each generalized tournament* $G \in \mathcal{T}_{n,m}^{g}$ *contains at least* $\mathcal{C}(n, m)$ *consistent triads* CT_{2a} *or* CT_3 *where*

$$\mathcal{C}(n, m) = \frac{1}{2} \left\lfloor \frac{m}{n} \right\rfloor \left(2m - n \left\lfloor \frac{m}{n} \right\rfloor - n \right). \tag{9.44}$$

Proof. The thesis follows from (Theorem 14 and 15). Theorem 14 determines the number of triads CT_{2a} or CT_3 for a vertex, while the second proves that the sum of triads CT_{2a} or CT_3 introduced by the vertices is minimal when their in-degrees are evenly distributed. Let us assume that in-degrees are evenly distributed.

Since there are m directed edges in G, then the sum of in-degrees of vertices is m. Thus, following the assumption about even distribution, every vertex needs to have at least $\left\lfloor \frac{m}{n} \right\rfloor$ incoming edges (when n is even, the in-degree of some vertices may be $\lfloor \frac{m}{n} \rfloor + 1$). Thus, in the our generalized tournament, there are p vertices with the in-degree $\lfloor \frac{m}{n} \rfloor$ and $n - p$ vertices with the in-degree $\lfloor \frac{m}{n} \rfloor + 1$. Following (Theorem 14), this graph has at least $\mathcal{C}(n,m)$ consistent triads, where

$$\mathcal{C}(n,m) = p \binom{\lfloor \frac{m}{n} \rfloor}{2} + (n-p) \binom{\lfloor \frac{m}{n} \rfloor + 1}{2}. \tag{9.45}$$

As the sum of in-degrees for all vertices is m, so

$$p \left\lfloor \frac{m}{n} \right\rfloor + (n-p)\left(\left\lfloor \frac{m}{n} \right\rfloor + 1\right) = m. \tag{9.46}$$

Thus,

$$p = n\left(\left\lfloor \frac{m}{n} \right\rfloor + 1\right) - m. \tag{9.47}$$

Hence (9.45) takes the form

$$\mathcal{C}(n,m) = \left(n \cdot \left(\left\lfloor \frac{m}{n} \right\rfloor + 1\right) - m\right) \cdot \binom{\lfloor \frac{m}{n} \rfloor}{2}$$
$$+ \left(m - n \cdot \left\lfloor \frac{m}{n} \right\rfloor\right) \cdot \binom{\lfloor \frac{m}{n} \rfloor + 1}{2} \tag{9.48}$$

which is equivalent to (9.44). $\qquad\square$

As every generalized tournament has $\binom{n}{3}$ triads, the above lemma results in the following observation:

Corollary 3. *Every $G \in \mathscr{T}_{n,m}^g$ contains at most*

$$\binom{n}{3} - \mathcal{C}(n,m) \tag{9.49}$$

inconsistent triads.

In particular, the above observation is evidence that the most inconsistent graph cannot have only directed edges.

Let $\mathcal{T}$ denote a set of all the triads in the generalized tournament and $\mathcal{T}_i$ denote a set of triads covered by $i = 0, \ldots, 3$ directed edges. The sum $\mathcal{T}_i \cup \mathcal{T}_j$ will be abbreviated as $\mathcal{T}_{i,j}$. Of course, $\mathcal{T} = \mathcal{T}_0 \cup \mathcal{T}_1 \cup \mathcal{T}_{2,3}$. Since $\mathcal{T}_i \cap \mathcal{T}_j = \emptyset$ for $i,j = 1, \ldots, n$ and $i \neq j$ then we may formulate the following quite obvious but useful conclusion.

Corollary 4. *For every $G \in \mathscr{T}_n^g$, it holds that*

$$\binom{n}{3} = |\mathcal{T}_0| + |\mathcal{T}_1| + |\mathcal{T}_{2,3}|. \tag{9.50}$$

Our current goal will be to determine the value $|\mathcal{T}_0|$. To this end, we will provide several quantitative estimates for the elements $|\mathcal{T}_0|$, $|\mathcal{T}_1|$ and $|\mathcal{T}_{2,3}|$ that allow us to construct the formula for $|\mathcal{T}_0|$.

Lemma 5. *For every $G \in \mathscr{T}_n^g$ where $G = (V, E_u, E_d)$, it is true that*

$$\sum_{a \in V} \binom{deg_{un}(a)}{2} = 3\,|\mathcal{T}_0| + |\mathcal{T}_1|\,. \tag{9.51}$$

Proof. Let the undirected edges $a_1 - a, \ldots, a_k - a$ be adjacent to some $a \in V$. As every pair of adjacent edges determines one triad, then there are $\binom{k}{2}$ triads that contain a. If $\{a_i, a_j\} \in E_u$ then the triad a, a_i, a_j belongs to $\mathcal{T}_0$, whereas if $(a_i, a_j) \in E_d$ then the triad a, a_i, a_j is in $\mathcal{T}_1$. In the sum $\sum_{a \in V} \binom{deg_{un}(a)}{2}$ every CT_0 triad is considered three times, as there are three vertices adjacent to the two undirected edges in the given triad. Similarly, the IT_1 triads are counted only once. $\square$

Let us consider the "minimal" version of the above lemma.

Lemma 6. *For $G \in \mathscr{T}_{n,m}^g$ where $G = (V, E_u, E_d)$, it holds that*

$$\mathcal{D}(n, m) \le 3\,|\mathcal{T}_0| + |\mathcal{T}_1|\,, \tag{9.52}$$

where

$$\mathcal{D}(n, m) = \frac{1}{2}\left(n - \left\lfloor \frac{2m}{n} \right\rfloor - 2\right)\left(n^2 + n\left(\left\lfloor \frac{2m}{n} \right\rfloor - 1\right) - 4m\right). \tag{9.53}$$

Proof. When is $\sum_{a \in V} \binom{deg_{un}(a)}{2}$ minimal? It is minimal when the un-degree for different vertices is similar. In practice, this postulate means that for two a_i and a_j the difference between $deg_{un}(a_i)$ and $deg_{un}(a_j)$ is at most 1. Based on (Def. 17) we can see $deg_{un}(a) = deg(a) - deg_{in}(a) - deg_{out}(a)$, and since $deg(a) = n - 1$, then

$$deg_{un}(a) = n - 1 - (deg_{in}(a) + deg_{out}(a))\,. \tag{9.54}$$

This means that the un-degrees of vertices are evenly distributed when directed edges are evenly distributed among the vertices. As every (directed) edge is adjacent to two vertices, then for m directed edges the sum of in-degrees and out-degrees is $2m$

$$\sum_{a \in V} deg_{in}(a) + deg_{out}(a) = 2m.$$

As the difference between the number of directed edges adjacent to different vertices can be at most 1 thus we may have two groups of vertices. One group where the sum of in-degrees and out-degrees is $\left\lfloor \frac{2m}{n} \right\rfloor$ and the other group where it is $\left\lfloor \frac{2m}{n} \right\rfloor + 1$. Let us denote the number of vertices in the first group by p and in the second group by $n-p$, where $p \leq n$. This leads to the following equation:

$$p \left\lfloor \frac{2m}{n} \right\rfloor + (n-p) \left(\left\lfloor \frac{2m}{n} \right\rfloor + 1 \right) = 2m. \tag{9.55}$$

By (9.54) we understand that in the graph that minimizes the expression $\sum_{a \in V} \binom{\deg_{un}(a)}{2}$ there are p vertices $a_1, \ldots, a_p$ for which $\deg_{un}(a_i) = n - 1 - \left\lfloor \frac{2m}{n} \right\rfloor$ and $1 \leq i \leq p$, and $n - p$ vertices $a_{p+1}, \ldots, a_n$ for which $\deg_{un}(a_j) = n - 2 - \left\lfloor \frac{2m}{n} \right\rfloor$ where $p + 1 \leq j \leq n$.

Thus, the lower bound of the sum $3|\mathcal{T}_0| + |\mathcal{T}_1|$ is given:

$$\mathcal{D}(n, m) \overset{df}{=} p \binom{n - 1 - \left\lfloor \frac{2m}{n} \right\rfloor}{2} + (n-p) \binom{n - 2 - \left\lfloor \frac{2m}{n} \right\rfloor}{2}. \tag{9.56}$$

By transforming (9.55) we obtain

$$p = n \left(\left\lfloor \frac{2m}{n} \right\rfloor + 1 \right) - 2m. \tag{9.57}$$

Hence, applying (9.57) to (9.56) leads to

$$\mathcal{D}(n, m) = \left(n \left(\left\lfloor \frac{2m}{n} \right\rfloor + 1 \right) - 2m \right) \binom{n - 1 - \left\lfloor \frac{2m}{n} \right\rfloor}{2}$$
$$+ \left(2m - n \left\lfloor \frac{2m}{n} \right\rfloor \right) \binom{n - 2 - \left\lfloor \frac{2m}{n} \right\rfloor}{2}. \tag{9.58}$$

The above expression can be simplified to

$$\mathcal{D}(n, m) = \frac{1}{2} \left(- \left\lfloor \frac{2m}{n} \right\rfloor + n - 2 \right) \left(n \left\lfloor \frac{2m}{n} \right\rfloor - 4m + (n-1)n \right), \tag{9.59}$$

which completes the proof of the theorem. □

In the above theorem, we considered the sum $\deg_{in}(a) + \deg_{out}(a)$ for $a \in V$. This sum allows us to estimate the lower bound for $|\mathcal{T}_{2,3}|$.

Lemma 7. *For every $G \in \mathcal{T}_n^g$ where $G = (V, E_u, E_d)$, it is true that*

$$\frac{1}{3} \sum_{a \in V} \binom{deg_{in}(a) + deg_{out}(a)}{2} \leq |\mathcal{T}_{2,3}|. \tag{9.60}$$

Proof. Let E be the set of directed edges adjacent to $a \in V$. Of course, $|E| = k = \deg_{in}(a) + \deg_{out}(a)$. Every two edges from E determine a triad containing a. There are $\binom{k}{2}$ such triads. Every triad is covered by two or three directed edges. While calculating the sum $\sum_{a \in V} \binom{\deg_{in}(a)+\deg_{out}(a)}{2}$ triads covered by two directed edges are considered once. The triads covered by three directed edges are considered three times. Thus, $\frac{1}{3} \sum_{a \in V} \binom{\deg_{in}(a)+\deg_{out}(a)}{2}$ limits $|\mathcal{T}_{2,3}|$ from below. $\qquad\square$

Once again, we can minimize the left side of (9.60) by assuming that the sum of in-degrees and out-degrees is evenly distributed among the vertices.

Lemma 8. *For each $G \in \mathcal{T}_{n,m}^g$ where $G = (V, E_u, E_d)$, it holds that*

$$\mathcal{E}(n, m) \leq |\mathcal{T}_{2,3}|, \tag{9.61}$$

where

$$\mathcal{E}(n, m) = \frac{1}{6} \left\lfloor \frac{2m}{n} \right\rfloor \left(4m - n \left(\left\lfloor \frac{2m}{n} \right\rfloor + 1 \right) \right). \tag{9.62}$$

Proof. The sum of input and output degrees of vertices in a generalized tournament with m directed edges is $2m$. In particular, for this graph it holds that (9.55). Thus, once again there should be p vertices adjacent to $\left\lfloor \frac{2m}{n} \right\rfloor$ directed edges, and $n - p$ vertices adjacent to $\left\lfloor \frac{2m}{n} \right\rfloor + 1$ directed edges. In other words:

$$p \binom{\left\lfloor \frac{2m}{n} \right\rfloor}{2} + (n - p) \binom{\left\lfloor \frac{2m}{n} \right\rfloor + 1}{2} \leq \binom{\deg_{in}(a) + \deg_{out}(a)}{2}.$$

Thus, our "candidate" for the lower bound obtains the form

$$\mathcal{E}(n, m) = \frac{1}{3} \left(p \binom{\left\lfloor \frac{2m}{n} \right\rfloor}{2} + (n - p) \binom{\left\lfloor \frac{2m}{n} \right\rfloor + 1}{2} \right).$$

Similarly as in Lemma 6 we use (9.57) and obtain

$$\mathcal{E}(n, m) = \frac{1}{3} \left\{ \left[n \left(\left\lfloor \frac{2m}{n} \right\rfloor + 1 \right) - 2m \right] \binom{\left\lfloor \frac{2m}{n} \right\rfloor}{2} + \left[n - \left(n \left(\left\lfloor \frac{2m}{n} \right\rfloor + 1 \right) - 2m \right) \right] \binom{\left\lfloor \frac{2m}{n} \right\rfloor + 1}{2} \right\}.$$

Hence,

$$\mathcal{E}(n, m) = \frac{1}{3} \left\{ \left(n \left\lfloor \frac{2m}{n} \right\rfloor + n - 2m \right) \binom{\left\lfloor \frac{2m}{n} \right\rfloor}{2} + \left(2m - n \left\lfloor \frac{2m}{n} \right\rfloor \right) \binom{\left\lfloor \frac{2m}{n} \right\rfloor + 1}{2} \right\},$$

which can be simplified to

$$\mathcal{E}(n, m) = \frac{1}{6} \left\lfloor \frac{2m}{n} \right\rfloor \left(4m - n \left\lfloor \frac{2m}{n} \right\rfloor - n \right). \tag{9.63}$$

$\qquad\square$

All the Lemmas and Corollaries allow us to estimate the lower bound for the number of consistent triads without directed edges.

Theorem 18. *For each $G \in \mathscr{T}_{n,m}^g$, it holds that*

$$\mathcal{F}(n,m) \le |\mathcal{T}_0|, \tag{9.64}$$

where

$$\mathcal{F}(n,m) = \frac{1}{2}\left(\mathcal{D}(n,m) + \mathcal{E}(n,m) - \binom{n}{3}\right). \tag{9.65}$$

Proof. Because of (Corollary 4)

$$\binom{n}{3} = |\mathcal{T}_0| + |\mathcal{T}_1| + |\mathcal{T}_{2,3}|,$$

and (Lemma 6)

$$\mathcal{D}(n,m) - 3|\mathcal{T}_0| \le |\mathcal{T}_1|.$$

we obtain the following estimation

$$\binom{n}{3} \ge |\mathcal{T}_0| + (\mathcal{D}(n,m) - 3|\mathcal{T}_0|) + |\mathcal{T}_{2,3}| = \mathcal{D}(n,m) + |\mathcal{T}_{2,3}| - 2|\mathcal{T}_0|.$$

As (Lemma 8) it holds that $\mathcal{E}(n,m) \le |\mathcal{T}_{2,3}|$, then it is true that

$$\binom{n}{3} \ge \mathcal{D}(n,m) + \mathcal{E}(n,m) - 2|\mathcal{T}_0|.$$

Thus,

$$|\mathcal{T}_0| \ge \frac{1}{2}\left(\mathcal{D}(n,m) + \mathcal{E}(n,m) - \binom{n}{3}\right).$$

which is the expected result. The above expression can also be rewritten in the following form:

$$|\mathcal{T}_0| \ge \frac{1}{6}\left((8m - 2n)\left\lfloor\frac{2m}{n}\right\rfloor - 2n\left\lfloor\frac{2m}{n}\right\rfloor^2 + (n-2)((n-1)n - 6m)\right). \tag{9.66}$$

$\square$

Since $|\mathcal{T}_0|$ is the number of consistent triads CT_0 then it cannot be negative. Thus, the inequality (9.65) can be re-written in the following form:

$$\max\{0, \lceil \mathcal{F}(n,m) \rceil\} \le |\mathcal{T}_0|. \tag{9.67}$$

At this point, we receive a second restriction for the number of consistent triads (the first is given by Theorem 17). Combining these two restrictions, we get the function

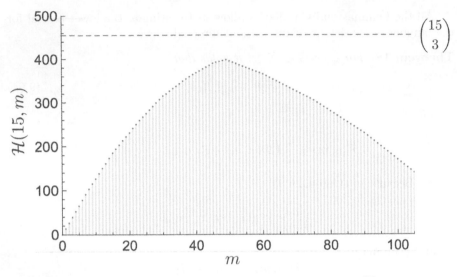

Figure 9.7: $\mathcal{H}(n, m)$ for $n = 15$ and $m = 0, \ldots, \binom{15}{2}$

$$\mathcal{G}(n, m) \overset{df}{=} \mathcal{C}(n, m) + \max\{0, \lceil \mathcal{F}(n, m) \rceil\}$$

that limits from below the number of consistent triads in $G \in \mathscr{T}_{n,m}^g$. This limitation immediately results in a limitation on the number of inconsistent triads. Indeed, the number of inconsistent triads in G is upper bounded by $\mathcal{H}(n, m)$ where:

$$\mathcal{H}(n, m) \overset{df}{=} \binom{n}{3} - \mathcal{G}(n, m). \tag{9.68}$$

The function $\mathcal{H}(n, m)$ limits the number of inconsistent triads in every generalized tournament with n vertices and m directed edges (Fig. 9.7). In particular, for any $n \geq 3$, its maximum limits from above the number of inconsistent triads in a maximally inconsistent generalized tournament with n vertices, i.e., for $G \in \overline{\mathscr{T}_n^g}$ and $n \geq 3$ it holds that:

$$|G|_i \leq \max_{0 \leq m \leq \binom{n}{2}} \mathcal{H}(n, m). \tag{9.69}$$

Our further efforts will be devoted to finding the graph $G \in \mathscr{T}_n^g$ for which it holds that $|G|_i = \max_{0 \leq m \leq \binom{n}{2}} \mathcal{H}(n, m)$. When we find it and prove that indeed it maximizes $\mathcal{H}$, this will mean that indeed $G \in \overline{\mathscr{T}_n^g}$, and we are able to determine the maximal number of inconsistent triads in a generalized tournament in n vertices.

9.2.3.2 The maximally inconsistent set of ordinal pairwise comparisons

In Figure 9.2, we can see that the maximally inconsistent tournament with n vertices is a graph in which input degrees are distributed as evenly as possible. It turns out that the most inconsistent generalized tournament has the form of two maximally inconsistent tournaments connected together by undirected edges in such a way that each vertex from one tournament is connected to each vertex from the second tournament (Def. 35). Furthermore, we will require both subgraphs to be the same size (of course, when the number of vertices of the generalized tournament is odd, the number of vertices in its subgraphs will differ by one). Let us propose our candidate for the most inconsistent generalized tournament.

Proposition 1. *The* double tournament $T = (V_1 \cup V_2, E_{d_1} \cup E_{d_2}, E_u)$ *is the maximally inconsistent generalized tournament if* (V_1, E_{d_1}) *and* (V_2, E_{d_2}) *are maximally inconsistent tournaments where* $|V_1| = \lfloor \frac{n}{2} \rfloor$ *and* $|V_2| = \lceil \frac{n}{2} \rceil$.

In Figure 9.8, we can see two examples of maximally inconsistent generalized tournaments with even G_{X^*}, and odd numbers of vertices G_{Y^*}. The ordinal PC matrices corresponding to G_{X^*} and G_{Y^*} are as follows:

$$X^* = \begin{pmatrix} 0 & -1 & 1 & 0 & 0 & 0 \\ 1 & 0 & -1 & 0 & 0 & 0 \\ -1 & 1 & 0 & 0 & 0 & 0 \\ 0 & 0 & 0 & 0 & -1 & 1 \\ 0 & 0 & 0 & 1 & 0 & -1 \\ 0 & 0 & 0 & -1 & 1 & 0 \end{pmatrix},$$

$$Y^* = \begin{pmatrix} 0 & -1 & 1 & 1 & 0 & 0 & 0 \\ 1 & 0 & -1 & -1 & 0 & 0 & 0 \\ -1 & 1 & 0 & -1 & 0 & 0 & 0 \\ -1 & 1 & 1 & 0 & 0 & 0 & 0 \\ 0 & 0 & 0 & 0 & 0 & 1 & -1 \\ 0 & 0 & 0 & 0 & -1 & 0 & 1 \\ 0 & 0 & 0 & 0 & 1 & -1 & 0 \end{pmatrix}.$$

The number of directed edges in the maximally inconsistent *dt-graph* is the sum of edges from both its sub-tournaments, i.e.,

$$\mathcal{X}(n) = \binom{\lfloor \frac{n}{2} \rfloor}{2} + \binom{\lceil \frac{n}{2} \rceil}{2}. \tag{9.70}$$

Depending on whether n is even or odd, it holds that $\mathcal{X}(2q) = q(q-1)$ if $n = 2q$ and $\mathcal{X}(2q+1) = q^2$ if $n = 2q + 1$, where $q \in \mathbb{N}_+$.

It is easy to observe that all the consistent triads in the dt-graph are in the two tournament subgraphs. Indeed, if a triad is not in a tournament, i.e.,

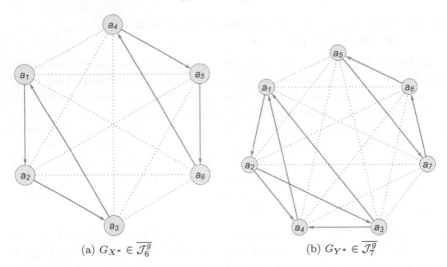

(a) $G_{X^*} \in \overline{\mathcal{J}_6^g}$ (b) $G_{Y^*} \in \overline{\mathcal{J}_7^g}$

Figure 9.8: Two examples of maximally inconsistent tournaments (undirected edges were dotted)

it is spanned between two tournaments, it must consist of one directed edge and two undirected edges. Thus, it must be inconsistent. It is easy to observe that in a double tournament all triads are covered by at least one directed edge. Thus, for every *dt-graph* it is true that $\max\{0, \lceil \mathcal{F}(n,m) \rceil\} = 0$. This does not guarantee, however, that $\mathcal{C}(n,m)$ – the second component of $\mathcal{G}$ – is also minimal.

Due to (Theorem 16), the maximal number of consistent triads in both parts of a maximally inconsistent generalized tournament are $\binom{\lfloor \frac{n}{2} \rfloor}{3} - \mathcal{I}\left(\lfloor \frac{n}{2} \rfloor\right)$ and $\binom{\lceil \frac{n}{2} \rceil}{3} - \mathcal{I}\left(\lceil \frac{n}{2} \rceil\right)$ correspondingly. As all the consistent triads in a double tournament are in its subgraphs, thus the number of inconsistent triads in the maximally inconsistent double tournament is given as (9.22).

The logic confirming the fact that indeed the graph as defined in (Proposition 1) is the maximally inconsistent dt-graph requires us to show that $\mathcal{H}$ gets maximal values when the number of directed edges is $\mathcal{X}(n)$, i.e., $\mathcal{Y}(n) = \mathcal{H}(n, \mathcal{X}(n)) \geq \mathcal{H}(n,m)$ for every $m = 1, \ldots, \binom{n}{2}$. Following [113], the reasoning is composed of four lemmas and based on a theorem. Because proofs of the lemmas are cumbersome, though not very conceptually complicated, they will not be given here. Interested readers can find them in [113].

Lemma 9. *For every $n \in \mathbb{N}_+, n \geq 3$ and $k \in \mathbb{N}_+$ it holds that:*

$$\mathcal{F}(n, \mathcal{X}(n)) = 0, \tag{9.71}$$

$$\mathcal{F}(n, \mathcal{X}(n) - k) \geq 1, \quad \text{where } 0 < k < \mathcal{X}(n), \tag{9.72}$$

$$\mathcal{F}(n, \mathcal{X}(n) + k) \leq 0, \quad \text{where } 0 < k \leq \binom{n}{2} - \mathcal{X}(n). \tag{9.73}$$

Lemma 10. *For every* $n \in \mathbb{N}_+, n \geq 3$ *the function* $\mathcal{C}$

1. *is constant and equals* $\mathcal{C}(n, m) = 0$ *for every* m *such that* $0 \leq m < n$, *and*

2. *is strictly increasing for every* $m \in \mathbb{N}_+$ *such that* $n \leq m \leq \binom{n}{2}$, *i.e.,*

$$\mathcal{C}(n, m+1) - \mathcal{C}(n, m) > 0. \tag{9.74}$$

Lemma 11. *For every* $n \in \mathbb{N}_+, n \geq 3$ *it holds that*

$$\mathcal{Y}(n) = \binom{n}{3} - \mathcal{C}(n, \mathcal{X}(n)). \tag{9.75}$$

Lemma 12. *For every* $n \in \mathbb{N}_+, n \geq 3$ *the function* $\mathcal{G}$ *is strictly decreasing for every* $m \in \mathbb{N}_+$ *such that* $1 \leq m \leq \mathcal{X}(n)$, *i.e.,*

$$\mathcal{G}(n, m) - \mathcal{G}(n, m+1) > 0 \quad \text{where } 1 \leq m < \mathcal{X}(n). \tag{9.76}$$

The Lemmas 9–12 allow us to prove the final theorem confirming that a double tournament composed of two possibly equal maximally inconsistent tournaments (Proposition 1) is indeed the maximally inconsistent generalized tournament. This theorem, in particular, justifies the value $\mathcal{Y}(n)$ and, as follows, the formula (9.23) defining the generalized Kendall–Babington Smith index.

Theorem 19. *For every double tournament* $G = (V_1 \cup V_2, E_{d_1} \cup E_{d_2}, E_u)$ *with* n *vertices where the tournaments* (V_1, E_{d_1}) *and* (V_2, E_{d_2}) *are maximally inconsistent and* $|V_1| = \lfloor \frac{n}{2} \rfloor$ *and* $|V_2| = \lceil \frac{n}{2} \rceil$, $n > 3$, *it is true that:*

1.
$$\mathcal{H}(n, \mathcal{X}(n)) = \max_{0 \leq m \leq \binom{n}{2}} \mathcal{H}(n, m), \tag{9.77}$$

2.
$$\mathcal{H}(n, \mathcal{X}(n)) = \mathcal{Y}(n) \tag{9.78}$$

Proof. As (9.68), then the first claim of the theorem is equivalent to

$$\mathcal{G}(n, \mathcal{X}(n)) = \min_{0 \leq m \leq \binom{n}{2}} \mathcal{G}(n, m) \tag{9.79}$$

Due to (Lemma 10) $\mathcal{C}$ does not decrease with respect to m. On the other hand (Lemma 9) $\mathcal{F}(n, \mathcal{X}(n) + k) \leq 0$ for every $0 < k \leq \binom{n}{2} - \mathcal{X}(n)$, which means that for every $m \geq \mathcal{X}(n)$ it holds that $\max\{0, \lceil \mathcal{F}(n, m) \rceil\} = 0$. Hence, for every $m \geq \mathcal{X}(n)$ there is the equivalence $\mathcal{G}(n,m) = \mathcal{C}(n,m)$. As $\mathcal{C}(n,m)$ is non-decreasing (Lemma 10) then

$$\mathcal{G}(n,\mathcal{X}(n)) \leq \mathcal{G}(n,\mathcal{X}(n)+1) \leq \ldots \leq \mathcal{G}(n, \binom{n}{2})) \qquad (9.80)$$

The above, together with (Lemma 12), i.e.,

$$\mathcal{G}(n,0) > \mathcal{G}(n,1) > \ldots > \mathcal{G}(n,\mathcal{X}(n)) \qquad (9.81)$$

implies

$$\mathcal{G}(n,\mathcal{X}(n)) = \min_{0 \leq m \leq \binom{n}{2}} \mathcal{G}(n,m) \qquad (9.82)$$

which completes the first part of the proof. The proof of the second claim easily follows from the fact that for every $m \geq \mathcal{X}(n)$ it holds that $\mathcal{G}(n,m) = \mathcal{C}(n,m)$. Thus, in particular

$$\mathcal{H}(n,\mathcal{X}(n)) = \binom{n}{3} - \mathcal{G}(n,\mathcal{X}(n)) = \binom{n}{3} - \mathcal{C}(n,\mathcal{X}(n)) \qquad (9.83)$$

which needs to be shown. $\qquad \square$

Chapter 10

Fuzzy AHP

In Section 2.7, we could see that fuzzy numbers are not actually numbers, and the operations on them are more complicated than ordinary ones. Hence, on one hand, it is easier to express preferences in the form of fuzzy numbers, but on the other, it is much more difficult to calculate and interpret a ranking. In the context of AHP, several different solutions can be found in the literature. All of the ones presented in this chapter are based on the triangular fuzzy numbers introduced in Section 2.7 and the logarithmic least square idea (compare with Section 3.3.3). *Laarhoven* and *Pedrycz's* solution is presented first [194]. Next, the optimal solution devised by *Ramik* and *Korviny* is shown [148]. We also look closer at the idea of creating a ranking of fuzzy numbers [174, 195] as well as creating an exact ranking based on fuzzy preferences [39].

10.1   Fuzzy LLSM

10.1.1   Direct approach

The idea of fuzzy LLSM as presented in [194] is based on the observation that the fuzzy priority vector candidate w should minimize the expression

$$\sum_{i<j} \sum_{k=1}^{\delta_{ij}} \left(\ln c_{ijk} - \ln w(a_i) + \ln w(a_j) \right)^2, \qquad (10.1)$$

where $c_{ijk}, w(a_i)$ and $w(a_j)$ are fuzzy numbers, and δ_{ij} denotes the number of estimates that we have for comparisons of the i-th and j-th alternatives. This is, in fact, direct mapping of LLSM as it is known for standard AHP (see Section 3.3.1) onto fuzzy numbers. By adopting $\widehat{c}_{ijk} = \ln c_{ijk}$ and $\widehat{w}(a_i) = \ln w(a)$ the expression (10.1) obtains the form

$$\sum_{i<j}\sum_{k=1}^{\delta_{ij}} (\widehat{c}_{ijk} - \widehat{w}(a_i) + \widehat{w}(a_j))^2.$$
(10.2)

By differentiating the above equation we get the linear equation system

$$\widehat{w}(a_i) \sum_{\substack{j=1 \\ j\neq i}}^{n} \delta_{ij} - \sum_{\substack{j=1 \\ j\neq i}}^{n} \widehat{w}(a_j)\delta_{ij} = \sum_{\substack{j=1 \\ j\neq i}}^{n}\sum_{k=1}^{\delta_{ij}} \widehat{c}_{ijk}, \quad \text{for } i = 1, 2, \ldots, n.$$
(10.3)

whose solution is the desired, but logarithmically transformed, priority vector. In the case of fuzzy numbers, both the priority vector w and PC matrix $C = [c_{ij}]$ are fuzzy numbers. Let us denote the elements of a priority vector using triangular fuzzy numbers as $w(a_i) = (w_l(a_i), w_m(a_i), w_u(a_i))$, and similarly elements of the PC matrix C as $c_{ij} = (l_{ij}, m_{ij}, u_{ij})$. Of course, $\widehat{w}(a_i) = (\widehat{w}_l(a_i), \widehat{w}_m(a_i), \widehat{w}_u(a_i))$ and $\widehat{c}_{ijk} = \left(\widehat{c}_{ijk}^l, \widehat{c}_{ijk}^m, \widehat{c}_{ijk}^u\right)$. Then, having in mind the operations (2.1)–(2.6), we obtain three sets of equations:

$$\widehat{w}_l(a_i) \sum_{\substack{j=1 \\ j\neq i}}^{n} \delta_{ij} - \sum_{\substack{j=1 \\ j\neq i}}^{n} \widehat{w}_u(a_j)\delta_{ij} = \sum_{\substack{j=1 \\ j\neq i}}^{n}\sum_{k=1}^{\delta_{ij}} \widehat{c}_{ijk}^l, \quad \text{for } i = 1, 2, \ldots, n,$$
(10.4)

$$\widehat{w}_m(a_i) \sum_{\substack{j=1 \\ j\neq i}}^{n} \delta_{ij} - \sum_{\substack{j=1 \\ j\neq i}}^{n} \widehat{w}_m(a_j)\delta_{ij} = \sum_{\substack{j=1 \\ j\neq i}}^{n}\sum_{k=1}^{\delta_{ij}} \widehat{c}_{ijk}^m, \quad \text{for } i = 1, 2, \ldots, n,$$
(10.5)

$$\widehat{w}_u(a_i) \sum_{\substack{j=1 \\ j\neq i}}^{n} \delta_{ij} - \sum_{\substack{j=1 \\ j\neq i}}^{n} \widehat{w}_l(a_j)\delta_{ij} = \sum_{\substack{j=1 \\ j\neq i}}^{n}\sum_{k=1}^{\delta_{ij}} \widehat{c}_{ijk}^u, \quad \text{for } i = 1, 2, \ldots, n.$$
(10.6)

The equations (10.4)–(10.6) are lineally dependent and sum up to 0. The authors propose the following formula as a solution of the above

$$\widehat{w}(a_i) = (\widehat{w}_l(a_i) + p_1, \widehat{w}_m(a_i) + p_2, \widehat{w}_u(a_i) + p_1).$$
(10.7)

The parameters p_1 and p_2 are chosen arbitrarily. It may turn out, however, that it is not possible to choose the parameters p_1 and p_2 such that

$$\widehat{w}_l(a_i) + p_1 \leq \widehat{w}_m(a_i) + p_2 \leq \widehat{w}_u(a_i) + p_1,$$
(10.8)

i.e., the value $\widehat{w}(a_i)$ may not be a triangular fuzzy number. In such a case, the authors suggest looking at (10.3) as to the undetermined system in the form

$$A\widehat{w} = \widehat{c},$$

where its minimum-norm solution is given as

$$A^T \left(AA^T\right)^{-1} \widehat{c}.$$

If A does not have negative entries, then $A^T \left(AA^T\right)^{-1} \widehat{c}$ is the desired candidate for the logarithmized priority vector, i.e., $\widehat{w}$. If not, the authors recommend taking exponentials of (10.7) then normalizing it. This leads to the priority vector in the form

$$w(a_i) = \left(\gamma_1 \cdot \exp \widehat{w}_l(a_i), \gamma_2 \cdot \exp \widehat{w}_m(a_i), \gamma_3 \cdot \exp \widehat{w}_u(a_i)\right),$$

where

$$\gamma_1 = \left(\sum_{i=1}^{n} \exp \widehat{w}_u(a_i)\right)^{-1}, \gamma_2 = \left(\sum_{i=1}^{n} \exp \widehat{w}_m(a_i)\right)^{-1},$$

$$\gamma_3 = \left(\sum_{i=1}^{n} \exp \widehat{w}_l(a_i)\right)^{-1}.$$

Example 17. *Let us consider the following fuzzy matrix* C

$$C = \begin{pmatrix} \begin{pmatrix} 1 & 1 & 1 \\ \frac{3}{2} & 2 & \frac{5}{2} \\ \frac{2}{3} & 1 & \frac{3}{2} \\ \frac{5}{2} & 3 & \frac{7}{2} \end{pmatrix} & \begin{pmatrix} \frac{2}{5} & \frac{1}{2} & \frac{2}{3} \\ 1 & 1 & 1 \\ \frac{2}{7} & \frac{1}{3} & \frac{2}{5} \\ \frac{2}{3} & 1 & \frac{3}{2} \end{pmatrix} & \begin{pmatrix} \frac{2}{3} & 1 & \frac{3}{2} \\ \frac{5}{2} & 3 & \frac{7}{2} \\ 1 & 1 & 1 \\ \frac{1}{3} & \frac{2}{5} & \frac{5}{2} \end{pmatrix} & \begin{pmatrix} \frac{2}{7} & \frac{1}{3} & \frac{2}{5} \\ \frac{2}{3} & 1 & \frac{3}{2} \\ \frac{2}{5} & \frac{5}{2} & 3 \\ 1 & 1 & 1 \end{pmatrix} \end{pmatrix}.$$

By solving the equation system[1] *(10.4)–(10.6), (10.8) with the additional constraint* $p_1, p_2 > 0$ *we obtain the following result*

$$\widehat{w} = \begin{pmatrix} (0 + p_1, 0 + p_2, 0.02569 + p_1) \\ (0.8727 + p_1, 0.895 + p_2, 0.8983 + p_1) \\ (-0.188 + p_1, 0.4023 + p_2, 0.5897 + p_1) \\ (0.3086 + p_1, 0.4935 + p_2, 1.0864 + p_1) \end{pmatrix},$$

where $p_1 = p_2 = 1$. *Thus,*

$$\widehat{w} = \begin{pmatrix} (1, 1, 1.02569) \\ (1.8727, 1.895, 1.8983) \\ (0.8119, 1.4023, 1.5897) \\ (1.3086, 1.4935, 2.0864) \end{pmatrix}.$$

---

[1]Solved using the FindInstance function in Wolfram Mathematica™ ver. 12.

By taking the exponentials, we obtain the final form of the priority vector:

$$w = \begin{pmatrix} (\ 2.71828 & 2.71828 & 2.78902\ ) \\ (\ 6.50585 & 6.6584 & 6.67517\ ) \\ (\ 2.25239 & 4.06478 & 4.90238\ ) \\ (\ 3.70126 & 4.45274 & 8.05589\ ) \end{pmatrix}.$$

We may also rescale the above vector so that the middle values sum up to 1, *i.e.,* $w_l(a_1) + \ldots + w_l(a_4) = 1.$ *This will help us compare the results with the next ranking method. In such a case, we obtain the following vector:*

$$w = \begin{pmatrix} (\ 0.151908 & 0.151908 & 0.155862\ ) \\ (\ 0.363573 & 0.372098 & 0.373035\ ) \\ (\ 0.125872 & 0.227156 & 0.273965\ ) \\ (\ 0.206841 & 0.248837 & 0.450196\ ) \end{pmatrix}. \qquad (10.9)$$

10.1.2 Optimal solution

Ramik and *Korviny* [148] formulate the fuzzy LLSM optimization constraint a bit differently[2]. According to their approach, the priority vector

$$w = \begin{pmatrix} w(a_1) \\ \vdots \\ w(a_n) \end{pmatrix} = \begin{pmatrix} (w_l(a_1), w_m(a_1), w_u(a_1)) \\ \vdots \\ (w_l(a_n), w_m(a_n), w_u(a_n)) \end{pmatrix}.$$

should minimize the expression

$$\sum_{i,j=1}^{n} \left\{ \left(\widehat{c}_{ij}^{\,l} - \widehat{w}_l(a_i) + \widehat{w}_l(a_j) \right)^2 + \right.$$
$$+ \left(\widehat{c}_{ij}^{\,m} - \widehat{w}_m(a_i) + \widehat{w}_m(a_j) \right)^2 + \qquad (10.10)$$
$$\left. + \left(\widehat{c}_{ij}^{\,u} - \widehat{w}_u(a_i) + \widehat{w}_u(a_j) \right)^2 \right\},$$

providing that

$$\sum_{i=1}^{n} w_m(a_i) = 1,$$

and

$$w_l(a_i) \leq w_m(a_i) \leq w_u(a_i) \quad \text{for} \quad i = 1, \ldots, n,$$

---

[2]It is worth noting that (10.10) differs from (10.2).

where $C = [c_{ij}]$ is a PC matrix composed of triangular fuzzy numbers in the form $c_{ij} = (c_{ij}^l, c_{ij}^m, c_{ij}^u)$. They prove [148] that the optimal solution of the above problem is given as:

$$w_l(a_k) = \zeta_{min} \cdot \frac{\left(\prod_{j=1}^n c_{kj}^l\right)^{1/n}}{\sum_{i=1}^n \left(\prod_{j=1}^n c_{ij}^m\right)^{1/n}}, \tag{10.11}$$

$$w_m(a_k) = \frac{\left(\prod_{j=1}^n c_{kj}^m\right)^{1/n}}{\sum_{i=1}^n \left(\prod_{j=1}^n c_{ij}^m\right)^{1/n}}, \tag{10.12}$$

$$w_u(a_k) = \zeta_{max} \cdot \frac{\left(\prod_{j=1}^n c_{kj}^u\right)^{1/n}}{\sum_{i=1}^n \left(\prod_{j=1}^n c_{ij}^m\right)^{1/n}}, \tag{10.13}$$

where

$$\zeta_{min} = \min_{i=1,\ldots,n} \left\{ \frac{\left(\prod_{j=1}^n c_{ij}^m\right)^{\frac{1}{n}}}{\left(\prod_{j=1}^n c_{ij}^l\right)^{\frac{1}{n}}} \right\},$$

and

$$\zeta_{max} = \max_{i=1,\ldots,n} \left\{ \frac{\left(\prod_{j=1}^n c_{ij}^m\right)^{\frac{1}{n}}}{\left(\prod_{j=1}^n c_{ij}^u\right)^{\frac{1}{n}}} \right\}.$$

Example 18. *Let us consider the same fuzzy matrix C as in the Example 17. It is easy to compute that this time (expressions: (10.11)–(10.13)), the fuzzy priority vector is:*

$$w = \left(\begin{array}{c} (\quad 0.151429 \quad 0.151908 \quad 0.155963 \quad) \\ (\quad 0.362425 \quad 0.372098 \quad 0.373277 \quad) \\ (\quad 0.151429 \quad 0.227156 \quad 0.227156 \quad) \\ (\quad 0.248837 \quad 0.248837 \quad 0.373277 \quad) \end{array} \right). \tag{10.14}$$

In particular, one may observe that $w_m(a_1) + \ldots + w_m(a_4) = 1$. Interestingly, the obtained result is slightly different than in the previous method (after rescaling).

10.2   Order of alternatives in the fuzzy ranking

Having fuzzy numbers at the input of the ranking procedure, we obtain a numerical ranking which is a fuzzy vector. However, the question arises as to how we should interpret this result. Who is the winner and who is the loser? Methods allowing the ranking of fuzzy numbers to be created are useful in finding the answer to this question. One way to solve this problem comes down to the conversion of the fuzzy number x into the corresponding real number $r(x)$ that has the meaning of a priority (weight) assigned to the given fuzzy number. Thus, having two fuzzy numbers x and y, we assume that $x \prec y$ (x precedes y) if $r(x) < r(y)$. Similarly, $x \sim y$ (x is just as preferable as y) if $r(x) = r(y)$. Another way to solve this task is to determine the preference relationship "$\prec$" between these numbers. In this case, this relationship will not necessarily be quantitative. However, it is enough to determine the winner.

There are many methods to calculate a ranking of fuzzy numbers [136, 174, 19, 10, 9]. Below are two simple methods to determine the ranking of triangular fuzzy numbers.

10.2.1   Graded mean method

One of the simplest methods for determining the ranking of triangular fuzzy numbers is the graded mean method [174]. In this approach, every triangular fuzzy number $x = (x_l, x_m, x_u)$ corresponds to a real number $r(x)$ defined as the graded mean of the smallest possible, the most likely, and the greatest possible values of x:

$$r(x) = \frac{x_l + 4x_m + x_u}{6}.$$

Example 19. *Lets compute the graded mean for the priority vectors computed in the Examples 17 and 18. For the priority vector (10.9), we obtain*

$$\begin{pmatrix} r\left(w(a_1)\right) \\ r\left(w(a_2)\right) \\ r\left(w(a_3)\right) \\ r\left(w(a_4)\right) \end{pmatrix} = \begin{pmatrix} 0.152567 \\ 0.370834 \\ 0.218077 \\ 0.275398 \end{pmatrix}.$$

Thus, the order of alternatives is a_2, a_4, a_3 and a_1. The same ranking method applied to (10.14) results in the following weights:

$$\begin{pmatrix} r\left(w(a_1)\right) \\ r\left(w(a_2)\right) \\ r\left(w(a_3)\right) \\ r\left(w(a_4)\right) \end{pmatrix} = \begin{pmatrix} 0.152504 \\ 0.370683 \\ 0.214535 \\ 0.269577 \end{pmatrix}.$$

Both results are very similar. In particular, there is no difference between them as to the order of alternatives. This time the winner is also a_2. The subsequent alternatives are a_4, a_3 and a_1.

10.2.2 Ranking of fuzzy numbers based on the relative preference relation

Another example of a method for determining the ranking between triangular fuzzy numbers can be found in the work of Wang [195]. According to this approach, there is a binary preference relation P with a membership function $\mu_p(x, y)$ representing the preference degree of x over y such that

- P is reciprocal as for every two fuzzy numbers x, y it holds that $\mu_p(x, y) = 1 - \mu_p(y, x)$,

- P is transitive as for every three fuzzy numbers x, y, z it holds that $\mu_p(x, y) \geq \frac{1}{2}$ and $\mu_p(y, z) \geq \frac{1}{2}$ then $\mu_p(x, z) \geq \frac{1}{2}$,

- P is fuzzy total ordering as P is reciprocal and transitive.

For two fuzzy numbers x and y it is said that x precedes y, i.e., $x \prec y$ if $\mu_p(x, y) > \frac{1}{2}$. Similarly, we would assume that x is just as preferable as y (denoted as $x \sim y$) if $\mu_p(x, y) = \frac{1}{2}$.

The membership function μ_p for the triangular fuzzy numbers $x = (x_l, x_m, x_u)$ and $y = (y_l, y_m, y_u)$ is defined as

$$\mu_p(x, y) = \frac{1}{2} \cdot \left(\frac{(x_l - y_u) + 2(x_m - y_m) + (x_u - y_l)}{2\,\|T\|} + 1 \right),$$

where

$$\|T\| = \begin{cases} TT & \text{when } t_l^- \geq t_u^- \\ TT + 2\left(t_u^- - t_l^+\right) & \text{otherwise} \end{cases},$$

and

$$TT = \frac{1}{2} \left(\left(t_l^+ - t_u^-\right) + 2\left(t_m^+ - t_m^-\right) + \left(t_u^- - t_l^-\right) \right),$$

$$t_l^+ = \max\{x_l, y_l\},$$

$$t_l^- = \min\{x_l, y_l\},$$

$$t_m^+ = \max\{x_m, y_m\},$$

$$t_m^- = \min\{x_m, y_m\},$$

$$t_u^+ = \max\{x_u, y_u\},$$

$$t_u^- = \min\{x_u, y_u\}.$$

Example 20. *Let us calculate an* $M = [m_{ij}]$ *matrix such that* $m_{ij} = \mu_p(w(a_i), w(a_j))$ *where* $w = [w(a_1), \ldots, w(a_4)]^T$ *is the ranking vector computed in the Example 17:*

$$M = \begin{pmatrix} 0.5 & 0 & 0.109335 & 0 \\ 1. & 0.5 & 1. & 0.814603 \\ 0.890665 & 0 & 0.5 & 0.235912 \\ 1. & 0.185397 & 0.764088 & 0.5 \end{pmatrix}.$$

The first row of M implies that $w(a_1) \sim w(a_1)$, $w(a_1) \prec w(a_2)$, $w(a_1) \prec w(a_3)$ *and* $w(a_1) \prec w(a_4)$. *The second row:* $w(a_3) \prec w(a_2)$ *and* $w(a_4) \prec w(a_2)$ *and finally* $w(a_3) \prec w(a_4)$. *Hence, we get the following total ordering:* $w(a_2) \succ w(a_4) \succ w(a_3) \succ w(a_1)$. *It is worth noting that this is exactly the same order of alternatives as the one obtained using the graded mean method (Example 19).*

10.3   Exact solution based on fuzzy preferences

The method proposed by *Chang* [39, 203] is an example of a two-in-one approach. It combines both: fuzzy pairwise comparisons on the input to the priority deriving procedure and the crisp (real) priorities on the output. It consists of three steps, of which the first one is the calculation of fuzzy priorities and the other two are the calculation of the ranking of these priorities.

Let $C = [c_{ij}]$ be the fuzzy PC matrix composed of triangular fuzzy numbers. In the first step we determine fuzzy priorities for all alternatives according to the formula:

$$w(a_i) = \frac{S_i}{S_1 \oplus \ldots \oplus S_n}, \quad \text{for } i = 1, \ldots, n,$$

where

$$S_i = c_{i1} \oplus c_{i2} \oplus \ldots \oplus c_{in}.$$

In the second step, the degree of possibility that $w(a_i)$ succeeds $w(a_j)$ (denoted $w(a_i) \geq w(a_j)$) needs to be determined. *Chang* proposes [39] defining this degree as

$$V\left(w(a_i) \geq w(a_j)\right) = \sup_{x \geq y} \left[\min\left(\mu_{w(a_i)}(x), \mu_{w(a_j)}(y)\right)\right],$$

where $\mu_{w(a_i)}(x)$ and $\mu_{w(a_j)}(y)$ are the membership functions of $w(a_i)$ and $w(a_j)$ correspondingly. In the case of triangular fuzzy numbers, the above formula boils down to [27]:

$$V\left(w(a_i) \geq w(a_j)\right) = \begin{cases} 1 & \text{if } w_m(a_i) \leq w_m(a_j) \\ \frac{w_l(a_i)-w_u(a_j)}{(w_m(a_j)-w_u(a_j))-(w_m(a_i)-w_u(a_i))} & \begin{aligned}&\text{if } w_m(a_i) \leq w_m(a_j) \\ &\text{and } w_l(a_j) < w_u(a_i)\end{aligned} \\ 0 & \text{otherwise} \end{cases}.$$

In particular, it holds that if $w_m(a_i) \leq w_m(a_j)$ then $V\left(w(a_i) \geq w(a_j)\right) = 1$ [203, p. 451]. One may observe that if the membership functions of $w(a_i)$ and $w(a_j)$ overlap and $w_l(a_i) \leq w_u(a_j)$ then $V\left(w(a_i) \geq w(a_j)\right)$ is the value of the point where both plots cross each other. In other words, there is such $x \in \mathbb{R}$ that $V\left(w(a_i) \geq w(a_j)\right) = \mu_{w(a_i)}(x) = \mu_{w(a_j)}(x)$.

In the third step of the procedure, we compute the degree of possibility that $w(a_i)$ succeeds all other fuzzy priorities. Hence,

$$\begin{aligned} V\left(w(a_i) \geq w(a_1), \ldots, w(a_n)\right) &= \\ = V\left(w(a_i) \geq w(a_1)\right) \text{ and} &\ldots \text{and } V\left(w(a_i) \geq w(a_n)\right) \\ = \min_{\substack{j=1,\ldots,n \\ i \neq j}} &\left\{V\left(w(a_i) \geq w(a_j)\right)\right\}. \end{aligned}$$

The above values form the desired real valued priority vector r. We therefore have

$$r = \left[r\left(w(a_1)\right), \ldots, r\left(w(a_n)\right)\right]^T,$$

where

$$r\left(w(a_i)\right) = V\left(w(a_i) \geq w(a_1), \ldots, w(a_n)\right).$$

Of course, r can be rescaled so that their entries sum up to 1. Hence, finally,

$$r_1 = \left[r_1\left(w(a_1)\right), \ldots, r_1\left(w(a_n)\right)\right]^T,$$

where

$$r_1\left(w(a_i)\right) = \frac{r\left(w(a_i)\right)}{\sum_{j=1}^{n} r\left(w(a_j)\right)}.$$

It is worth noting that *Chang's* method can be easily used for group decision making. For a group of k experts, each of whom provides a fuzzy PC matrix $C_q = [c_{ij}^{(q)}]$ for $q = 1, \ldots, k$, it is enough to adopt $c_{ij} = \frac{1}{k}\left(c_{ij}^{(1)} \oplus \ldots \oplus c_{ij}^{(k)}\right)$ [203, p. 453]. Then, taking $C = [c_{ij}]$ as the input matrix, we must apply the above three calculation steps.

Example 21. *Let us consider the well known matrix C introduced in the Example 17. One may easily calculate that*

$$\begin{pmatrix} S_1 \\ S_2 \\ S_3 \\ S_4 \end{pmatrix} = \begin{pmatrix} \left(\frac{247}{105}, \frac{17}{6}, \frac{107}{30}\right) \\ \left(\frac{17}{3}, 7, \frac{17}{2}\right) \\ \left(\frac{247}{105}, \frac{29}{6}, \frac{59}{10}\right) \\ \left(\frac{9}{2}, \frac{27}{5}, \frac{17}{2}\right) \end{pmatrix}.$$

Hence,

$$S_1 \oplus \ldots \oplus S_n = \left(\frac{1041}{70}, \frac{301}{15}, \frac{397}{15}\right),$$

and as follows

$$(S_1 \oplus \ldots \oplus S_n)^{-1} = \left(\frac{15}{397}, \frac{15}{301}, \frac{70}{1041}\right).$$

This allows us to compute the fuzzy ranking vector

$$\begin{pmatrix} w(a_1) \\ w(a_2) \\ w(a_3) \\ w(a_4) \end{pmatrix} = \begin{pmatrix} \left(\frac{247}{2779}, \frac{85}{602}, \frac{749}{3123}\right) \\ \left(\frac{85}{397}, \frac{15}{43}, \frac{595}{1041}\right) \\ \left(\frac{247}{2779}, \frac{145}{602}, \frac{413}{1041}\right) \\ \left(\frac{135}{794}, \frac{81}{301}, \frac{595}{1041}\right) \end{pmatrix}.$$

The degrees of possibility that the fuzzy priority of the *i*-th alternative is greater than the fuzzy priority of the *j*-th alternative can be shown in the form of the matrix $V = [v_{ij}]$ where $v_{ij} = V\left(w(a_i) \geq w(a_j)\right)$. For the considered case it is:

$$V = \begin{pmatrix} - & 0.1102 & 0.602 & 0.353 \\ 1. & - & 1. & 1. \\ 1. & 0.628 & - & 0.889 \\ 1. & 0.817 & 1. & - \end{pmatrix}.$$

Finally, the r vector is given as

$$r = [0.11, 1., 0.628, 0.817]^T,$$

and after rescaling

$$r_1 = [0.043, 0.391, 0.245, 0.319]^T.$$

Once again, the results of the ranking indicate that the best alternative is a_2, then a_4, a_3 and a_1.

It is worth noting that all the considered methods (Examples: 19, 20 and 21) lead to the identical ordinal rank. However, the numerical values computed may differ (see Examples 19 and 21).

10.4   Other methods and discussion

The above mentioned techniques of using fuzzy numbers to calculate priorities in AHP do not exhaust all the possible ways in which the fuzzy sets theory can support the pairwise comparison method. In this context, it is worth mentioning the approach proposed by *Mikhailov* [130]. The author proposes the transformation of fuzzy pairwise comparisons into a series of α-cuts, then finding the crisp values corresponding to each interval set of pairwise comparisons. Aggregation of the results provides the final crisp priorities of alternatives. Group decision-making was also possible [131]. Modification of the *Van Laarhoven* and *Pedrycz* approach [194] has been proposed by *Kwiesielewicz* [122]. Starting from the same premises (fuzzy equations), he proposed a different solving method allowing solution indeterminism to be avoided. Research on fuzzy AHP conducted by *Yuen* resulted in the definition of the Primitive Cognitive Network Process [198, 200, 199].

Fuzzy AHP has also sparked a number of discussions about the legitimacy of using fuzzy numbers to create and compute rankings based on pairwise comparisons [27, p. 63]. Particularly interesting seems to be the opinion formed by *Saaty* that Fuzzy AHP adds an additional but unnecessary level of fuzziness which may affect the final outcome. Despite this complaint, *Ramík* and *Korviny* argue that Fuzzy AHP is just a generalization of the classical approach and, as such, it *"cannot be worse than the crisp model"* [148]. Undoubtedly, the definitive answer to the question about the legitimacy of using Fuzzy AHP requires further research and experimentation.

Chapter 11

Heuristic Rating Estimation

11.1 The idea of HRE

The pairwise comparison method underlying AHP assumes that the actual priorities of alternatives are unknown and must be determined based on their mutual comparisons. If we look at the prioritization process in an algorithmic way (Fig. 1.2) we can say that the input to the priority deriving procedure is the appropriate PC matrix (Chapter 2) and the output is the numerical ranking. However, the assumption that the priorities of all considered alternatives are a priori not known is not necessarily always met. In a situation where for some alternatives there is already a ranking, it is reasonable to include existing results in the new ranking instead of calculating it from the very beginning. An existing ranking can be the result of expert work. For example, let us try to imagine a situation in which a group of experts created a ranking of five grant applications. However, after the results were announced, it turned out (e.g., due to a court judgment) that two more applications should be included in the ranking. In this situation, the jury, without changing the results of the first ranking, must prepare a second one taking into account the additional alternatives. The initial ranking may also have its source in real data being the result of physical, economic, financial or sociological measurements. Let's suppose, knowing the popularity of several selected websites (by number of visits), we want to determine the presumed popularity of several new websites on similar topics. Based on the existing (natural) popularity ranking, it's enough to create a ranking extended by the additional, new websites.

Numerical results of this ranking will simply be the values of the expected popularity of the new solutions.

The Heuristic Rating Estimation (HRE) method adapts the priority deriving procedures known from AHP to the modified decision model [108, 115, 112]. It is therefore a certain extension of the AHP method. Unlike AHP, a set of alternatives is divided into two subsets: alternatives for which the ranking has yet to be determined and already ranked (known, reference) alternatives. We will denote the set of the former as A_K and the subset of unknown alternatives as A_U. We assume, without loss of generality, that $A_U = \{a_1, \ldots, a_k\}$ and $A_K = \{a_{k+1}, \ldots, a_n\}$. The set of alternatives (see Sec. 1.2.2) is given as $A = A_U \cup A_K$ where $A_U \cap A_K = \varnothing$. HRE gets on the input the existing ranking of "known" alternatives $w(a_{k+1}), \ldots, w(a_n)$ and the results of pairwise comparisons for all alternatives except for the comparison of A_K elements with themselves. On the output, it produces the ranking of "unknown" alternatives $w(a_1), \ldots, w(a_k)$ which is compatible with the initial set of weights for $a_{k+1}, \ldots, a_n$ (Fig. 11.1). Two HRE methods have been defined [108, 115]: additive (AHRE) and geometric (GHRE). The difference between them is similar to that between EVM and GMM. In the first, the individual priorities of the alternatives are approximated by the weighted arithmetic mean of the other alternatives while, in the second, by using the weighted geometric mean.

11.2 Additive HRE

At the core of EVM (Chapter 3.1) lies the observation that the individual comparison c_{ij} should approximate the ratio $w(a_i)/w(a_j)$ (3.2). When the PC matrix is consistent there is a guarantee that $w(a_i)/w(a_j) = c_{ij}$ for $i, j = 1, \ldots, n$ (Theorem 4). In the case of AHRE, we can follow the same principle. Hence, similarly as for EVM (3.4) we may approximate $w(a_i)$ by $c_{ij}w(a_j)$, i.e.,

$$w(a_i) \approx c_{ij}w(a_j).$$

Similarly as before, we demand $w(a_i)$ be the mean of all its approximations. This request leads to:

$$w(a_i) = \frac{1}{n-1} \sum_{\substack{j=1 \\ i \neq j}} c_{ij}w(a_j), \qquad (11.1)$$

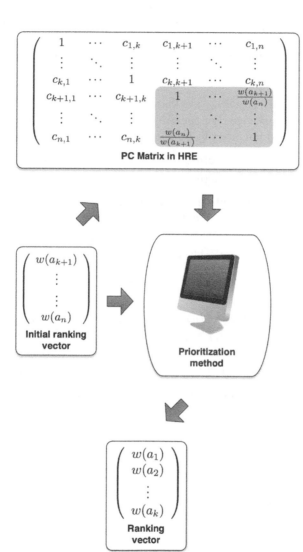

Figure 11.1: Prioritization procedure in the Heuristic Rating Estimation method

for $i, j = 1, \ldots, n$. The above postulate can be easily written in the form of the following equation system:

$$
\begin{aligned}
w(a_1) &= \tfrac{1}{n-1}(c_{1,2}w(a_2) + \ldots\ldots\ldots + c_{1,n}w(a_n)) \\
w(a_2) &= \tfrac{1}{n-1}(c_{2,1}w(a_1) + c_{2,3}w(a_3) + \ldots\ldots + c_{2,n}w(a_n)) \\
&\qquad\ldots\ldots\ldots\ldots\ldots\ldots\ldots\ldots\ldots \\
w(a_k) &= \tfrac{1}{n-1}(c_{k,1}w(a_1) + \ldots + c_{k,k-1}w(a_{k-1}) + c_{k,k+1}w(a_{k+1}) + \\
&\qquad\qquad\qquad + c_{k,n}w(a_n))
\end{aligned}
$$

$$(11.2)$$

The values $w(a_{k+1}), \ldots, w(a_n)$ are known and constant ($a_{k+1}, \ldots, a_n$ belongs to A_K known alternatives). For the same reason, all the comparisons c_{ij} where $i, j = k+1, \ldots n$ do not need to be determined by experts. Their values are:

$$
c_{k+1,k+2} = \frac{w(a_{k+1})}{w(a_{k+2})}, \ldots, c_{n,n-1} = \frac{w(a_n)}{w(a_{n-1})}.
$$

Therefore, all the components in the row composed of known elements can be summed up together. Let us denote:

$$
b_j = \frac{1}{n-1}c_{j,k+1}w(a_{k+1}) + \ldots + \frac{1}{n-1}c_{j,n}w(a_n).
$$

The above linear equations system (11.2) can be rewritten as:

$$
\begin{aligned}
w(a_1) &= \tfrac{1}{n-1}c_{1,2}w(a_2) + \ldots\ldots\ldots + \tfrac{1}{n-1}c_{k,1}w(a_k) + b_1 \\
w(a_2) &= \tfrac{1}{n-1}c_{2,1}w(a_1) + \tfrac{1}{n-1}c_{2,3}w(a_3) + \ldots + \tfrac{1}{n-1}c_{2,k}w(a_k) + b_2 \\
&\qquad\ldots\ldots\ldots\ldots\ldots\ldots\ldots\ldots\ldots\ldots \\
w(a_k) &= \tfrac{1}{n-1}c_{k,1}w(a_1) + \ldots\ldots + \tfrac{1}{n-1}c_{k,k-1}w(a_{k-1}) + b_k
\end{aligned}
$$

The matrix form of the above equation system is given as

$$\overline{C}w = b, \qquad (11.3)$$

where

$$
\overline{C} = \begin{pmatrix}
1 & -\tfrac{1}{n-1}c_{1,2} & \cdots & & -\tfrac{1}{n-1}c_{1,k} \\
-\tfrac{1}{n-1}c_{2,1} & 1 & \cdots & & -\tfrac{1}{n-1}c_{2,k} \\
\vdots & \vdots & \vdots & \vdots & \vdots \\
-\tfrac{1}{n-1}c_{k-1,1} & \cdots & & \ddots & -\tfrac{1}{n-1}c_{k-1,k} \\
-\tfrac{1}{n-1}c_{k,1} & \cdots & & -\tfrac{1}{n-1}c_{k,k-1} & 1
\end{pmatrix}, \qquad (11.4)
$$

the vector of constant terms is

$$
b = \begin{pmatrix}
\tfrac{1}{n-1}c_{1,k+1}w(a_{k+1}) + \ldots + \tfrac{1}{n-1}c_{1,n}w(a_n) \\
\tfrac{1}{n-1}c_{2,k+1}w(a_{k+1}) + \ldots + \tfrac{1}{n-1}c_{2,n}\mu(a_n) \\
\vdots \\
\tfrac{1}{n-1}c_{k,k+1}w(a_{k+1}) + \ldots + \tfrac{1}{n-1}c_{k,n}w(a_n)
\end{pmatrix}, \qquad (11.5)
$$

and values that need to be determined are denoted as:

$$w = \begin{pmatrix} w(a_1) \\ \vdots \\ \vdots \\ w(a_k) \end{pmatrix}. \tag{11.6}$$

After calculating the vector w, it can be completed with the reference elements:

$$w = \begin{pmatrix} w(a_1) \\ \vdots \\ w(a_k) \\ w(a_{k+1}) \\ \vdots \\ w(a_n) \end{pmatrix}.$$

In this way, we get a full ranking for all n alternatives. If the purpose of the ranking is only to rank individual alternatives (e.g., sort them out) and the weight does not have any other interpretation, then we may rescale the completed vector. In this case, the final ranking is given as:

$$w = \begin{pmatrix} \alpha \cdot w(a_1) \\ \vdots \\ \alpha \cdot w(a_k) \\ \alpha \cdot w(a_{k+1}) \\ \vdots \\ \alpha \cdot w(a_n) \end{pmatrix}, \quad \text{where } \alpha = \sum_{i=1}^{n} w(a_i).$$

Example 22. *Let $a_1, \ldots, a_5$ represent different universities, where two of them a_2 and a_3 are the reference alternatives for which the priorities have been determined in advance by the state evaluation committee[1]. These priorities are accordingly $w(a_2) = 5$ and $w(a_3) = 7$. The experts' pairwise assessment of the remaining alternatives a_1, a_4 and a_5 with respect to some criterion w allowed the formulation of the following PC matrix:*

$$C = \begin{pmatrix} 1 & \frac{3}{5} & \frac{4}{7} & \frac{3}{8} & \frac{4}{9} \\ \frac{5}{3} & 1 & \boxed{\frac{5}{7}} & \frac{5}{2} & \frac{10}{3} \\ \frac{7}{4} & \boxed{\frac{7}{5}} & 1 & \frac{5}{2} & 4 \\ \frac{8}{3} & \frac{2}{5} & \frac{2}{5} & 1 & \frac{4}{3} \\ \frac{9}{4} & \frac{3}{10} & \frac{1}{4} & \frac{3}{4} & 1 \end{pmatrix},$$

---

[1]Indeed, in 2013 in Poland, the algorithm for the evaluation of scientific units assumed the existence of reference units [101, 108].

where $c_{2,3} = 1/c_{3,2} = w(a_2)/w(a_3) = 5/7$. *The auxiliary matrix* $\overline{C}$ *gets the form*

$$
\overline{C} = \begin{bmatrix} 1 & -\frac{1}{n-1}c_{1,4} & -\frac{1}{n-1}c_{1,5} \\ -\frac{1}{n-1}c_{4,1} & 1 & -\frac{1}{n-1}c_{4,5} \\ -\frac{1}{n-1}c_{5,1} & -\frac{1}{n-1}c_{5,4} & 1 \end{bmatrix},
$$

and the constant term vector is

$$
b = \begin{bmatrix} \frac{1}{n-1}c_{1,2}w(c_2) + \frac{1}{n-1}c_{1,3}w(c_3) \\ \frac{1}{n-1}c_{4,2}w(c_2) + \frac{1}{n-1}c_{4,3}w(c_3) \\ \frac{1}{n-1}c_{5,2}w(c_2) + \frac{1}{n-1}c_{5,3}w(c_3) \end{bmatrix}.
$$

Using values from C, *we can write down the linear equation system (11.3) as*

$$
\begin{pmatrix} 1 & -\frac{3}{32} & -\frac{1}{9} \\ -\frac{2}{3} & 1 & -\frac{1}{3} \\ -\frac{9}{16} & -\frac{3}{16} & 1 \end{pmatrix} \begin{pmatrix} w(a_1) \\ w(a_4) \\ w(a_5) \end{pmatrix} = \begin{pmatrix} \frac{7}{4} \\ \frac{6}{5} \\ \frac{13}{16} \end{pmatrix}.
$$

As a solution we obtain

$$
\begin{pmatrix} w(a_1) \\ w(a_4) \\ w(a_5) \end{pmatrix} = \begin{pmatrix} \frac{43633}{17995} \\ \frac{40792}{10797} \\ \frac{51912}{17995} \end{pmatrix} = \begin{pmatrix} 2.424 \\ 3.778 \\ 2.885 \end{pmatrix}.
$$

The obtained weights are compatible with the initial ranking for a_2 *and* a_3. *Hence, we can safely complete the above vector with the reference values. The ranking of all the considered alternatives is therefore:*

$$
\begin{pmatrix} w(a_1) \\ w(a_2) \\ w(a_3) \\ w(a_4) \\ w(a_5) \end{pmatrix} = \begin{pmatrix} 2.424 \\ 5 \\ 7 \\ 3.778 \\ 2.885 \end{pmatrix}.
$$

As in AHP, the results obtained can be scaled so that their sum is 1. In such a case, the total ranking is

$$
w = [0.115, 0.237, 0.332, 0.1792, 0.1368]^T.
$$

11.3   Existence of a solution for additive HRE

In the previous section, we saw how we can easily add new alternatives to the existing pairwise comparison-based ranking. To this end, we had to solve a system of linear equations (11.3). In principle, however, this matrix equation does not need to have a solution. Fortunately, very often such a solution exists. In this section, following [112], we will formulate the sufficient condition for the existence of the solution and show its relationship with the inconsistency of the PC matrix. To this end, let us recall some basic information about a special class of matrices called M-matrices [146].

11.3.1 M-matrices

Let us denote $\mathcal{M}_{\mathbb{R}}(n)$ as the set of $n \times n$ matrices over $\mathbb{R}$, $\mathcal{M}_Z(n)$ as the set of all $M = (m_{ij}) \in \mathcal{M}_{\mathbb{R}}(n)$ with $m_{ij} \leq 0$ if $i \neq j$ and $1 \leq i, j \leq n$. Moreover, for every matrix $M \in \mathcal{M}_{\mathbb{R}}(n)$ and vector $r \in \mathbb{R}^n$ we assume that $M \geq 0$ and $r \geq 0$ mean that each entry of both M and r is non-negative and neither M nor r equals 0. The principal eigenvalue of M (the spectral radius of M) is denoted as $\rho(M)$ and equals $\rho(M) = \max\{|\lambda| : \det(\lambda I - M) = 0\}$.

Definition 37. *An $n \times n$ matrix that can be expressed in the form $M = sI - B$ where $B = [b_{ij}]$ with $b_{ij} \geq 0$ for $1 \leq i, j \leq n$, and $s \geq \rho(B)$, is called an M-matrix.*

The main areas of application of M-matrices are social, physical and biological sciences [143]. For example, they are used in constructing the Lotka-Volterra model from ecology, Leontief input-output models in economics, linear complementarity problems, e.g., the problem of finding an equilibrium point of a bimatrix game (game theory), Markov chains and others [16]. Due to their numerous applications, *M-matrices* have been of interest to researchers for a long time and many of their properties are known. Following *Plemmons* [143], the selected properties of M-matrices are recalled below in the form of the following theorem.

Theorem 20. *For every $M \in \mathcal{M}_Z(n)$, the following conditions are equivalent:*

1. *M is inverse positive. That is, M^{-1} exists and $M^{-1} \geq 0$,*

2. *M is semi-positive. That is, there is a vector $x > 0$ with $Mx > 0$,*

3. *There is a positive diagonal matrix D such that MD has all positive row sums,*

4. *M is a non-singular M-matrix.*

Note that for a non-singular M also M^{-1} is non-singular. Therefore, the solution of $Mw = r$ is $M^{-1}r$. Since $r > 0$ and M – M-matrix, then thanks to the theorem above it holds that $M^{-1} \geq 0$, and w must also be strictly positive, i.e., $w = M^{-1}r > 0$.

11.3.2 Sufficient condition for the existence of a solution

It is easy to note that PC matrices are positive, i.e., for the given PC matrix $C = [c_{ij}]$ it holds that $c_{ij} > 0$ for $i, j = 1, \ldots, n$. This is the reason that the auxiliary matrix $\overline{C}$ has a positive diagonal while outside the diagonal is negative, i.e., $\overline{C} \in \mathcal{M}_Z(n)$. In other words, following Theorem 20, if $\overline{C}$ meets any of the explicitly indicated conditions then it is an M-matrix. Hence, based on the same Theorem, $\overline{C}$ is non-singular (11.3) and it has an admissible, i.e., real and positive, solution.

Theorem 21. *The matrix equation $\overline{C}w = b$ (see 11.3) has one strictly positive solution if*

$$K(C) < L(n, r), \tag{11.7}$$

where C is an $n \times n$ PC matrix, $K(C)$ is the Koczkodaj inconsistency index of C (6.24), $n = |A_U \cup A_K|$ is the total number of alternatives , $r = |A_K|$ – is the number of reference alternatives and $0 < r \leq n - 2$ and

$$L(n, r) \overset{df}{=} 1 - \frac{1 + \sqrt{1 + 4(n-1)(n-r-2)}}{2(n-1)}.$$

Proof. Following (6.23, 6.24), Koczkodaj's *inconsistency index* $K(C)$, abbreviated as K, equals $K_{p,q,r}(C)$ where c_{pq}, c_{qr} and c_{pr} correspond to the most inconsistent triad in C. Thus, the local inconsistency for any other triad cannot be greater than that. Hence, for any c_{ik}, c_{kj}, c_{ij} it must hold that:

$$K \geq K_{i,j,k}(C) = \min \left\{ \left| 1 - \frac{c_{ij}}{c_{ik}c_{kj}} \right|, \left| 1 - \frac{c_{ik}c_{kj}}{c_{ij}} \right| \right\}. \tag{11.8}$$

Let us denote

$$\alpha \overset{df}{=} 1 - K. \tag{11.9}$$

It can be shown[2] (Lemma 2) that

$$\alpha \cdot c_{ik}c_{kj} \leq c_{ij} \leq \frac{1}{\alpha} c_{ik}c_{kj}, \tag{11.10}$$

for every i, j, k such that $1 \leq i, j, k \leq n$. The above inequality allows the parametric equation $c_{ij} = t \cdot c_{ik}c_{kj}$ to be written where $\alpha \leq t \leq \frac{1}{\alpha}$. With the help of this equation, the auxiliary matrix $\overline{C}$ can be written as:

$$\overline{C} = \begin{bmatrix} t_{1,1}c_{1,k}c_{k,1} & \cdots & -\frac{t_{1,k-1}c_{1,k}c_{k,k-1}}{n-1} & -\frac{c_{1,k}}{n-1} \\ \vdots & \vdots & \vdots & \vdots \\ -\frac{t_{k-1,1}c_{k-1,k}c_{k,1}}{n-1} & \ddots & t_{k-1,k-1}c_{k-1,k}c_{k,k-1} & -\frac{c_{k-1,k}}{n-1} \\ -\frac{t_{k,1}c_{k,1}}{n-1} & \cdots & -\frac{t_{k,k-1}c_{k,k-1}}{n-1} & 1 \end{bmatrix}, \tag{11.11}$$

---

[2]This can also be proven by contradiction. For example, let us assume that $c_{ij} < \alpha c_{ik}c_{kj}$ and a_i, a_j and a_k is the most inconsistent triad in C. In such a case $c_{ij} < (1 - K)c_{ik}c_{kj}$.

where $\alpha \leq t_{ij} \leq \frac{1}{\alpha}$, for i, j such that $1 \leq i, j \leq k - 1$. Going one step further, you will notice that $\overline{C}$ can be written as the product $\overline{C} = EF$ where:

$$E = \begin{bmatrix} t_{1,1}c_{1,k} & \cdots & -\frac{t_{1,k-1}c_{k-1,k}}{n-1} & -\frac{c_{1,k}}{n-1} \\ \vdots & \ddots & \vdots & \vdots \\ -\frac{t_{k-1,1}c_{k-1,k}}{n-1} & \vdots & t_{k-1,k-1}c_{k-1,k} & -\frac{c_{k-1,k}}{n-1} \\ -\frac{t_{k,1}}{n-1} & \cdots & -\frac{t_{k,k-1}}{n-1} & 1 \end{bmatrix} \tag{11.12}$$

and

$$F = \begin{bmatrix} c_{k,1} & 0 & \cdots & 0 \\ 0 & \ddots & \cdots & 0 \\ \vdots & \vdots & c_{k,k-1} & \vdots \\ 0 & \cdots & \cdots & 1 \end{bmatrix}. \tag{11.13}$$

Since all the parameters t_{ij} and matrix elements c_{ij} are strictly positive then $E \in \mathcal{M}_Z(n)$. Therefore, due to Theorem 20, with $D \stackrel{df}{=} I$, E is a non-singular *M-matrix* if and only if the sums of its rows are positive. Let us consider $(n-1)E$ which leads us to the system of the following inequalities:

$$\begin{aligned} c_{1,k}(n-1)t_{1,1} - c_{1,k}(t_{1,2} + t_{1,3} + \ldots + t_{1,k-1} + 1) &> 0 \\ c_{2,k}(n-1)t_{2,2} - c_{2,k}(t_{2,1} + t_{2,3} + \ldots + t_{2,k-1} + 1) &> 0 \\ \vdots \\ (n-1) - (t_{k,1} + t_{k,2} + \ldots\ldots + t_{k,k-1}) &> 0 \end{aligned} \tag{11.14}$$

Since (11.10) $\alpha \leq t_{ij} \leq \frac{1}{\alpha}$. Thus, the above inequalities hold if the following two constraints are met [112, p. 115]:

$$(n-1)\alpha > (\underbrace{\frac{1}{\alpha} + \ldots + \frac{1}{\alpha}}_{n-r-2} + 1) \quad \text{and} \quad (n-1) > (\underbrace{\frac{1}{\alpha} + \ldots + \frac{1}{\alpha}}_{n-r-1}), \tag{11.15}$$

where $r = n - k$ is the number of known alternatives $|A_K|$. Hence, E is *an M-matrix* if the following two conditions are met:

$$f(\alpha) > 0, \quad \text{where } f(\alpha) \stackrel{df}{=} (n-1)\alpha^2 - \alpha - (n-r-2) \tag{11.16}$$

and

$$g(\alpha) > 0, \quad \text{where } g(\alpha) \stackrel{df}{=} (n-1)\alpha - (n-r-1). \tag{11.17}$$

Solving $f(\alpha) = 0$ and choosing the larger root [3] we obtain:

$$K(C) < L(n,r) = 1 - \frac{1 + \sqrt{1 + 4(n-1)(n-r-2)}}{2(n-1)}. \tag{11.18}$$

---

[3]The smaller root of the two is negative for any $n = 3, 4 \ldots$ and $0 < r \leq n - 2$.

$K(C)$	$r = 1$	$r = 2$	$r = 3$	$r = 4$	$r = 5$	$r = 6$	$r = 7$	$r = 8$
$n = 3$	0.5	–	–	–	–	–	–	–
$n = 4$	0.232	0.667	–	–	–	–	–	–
$n = 5$	0.157	0.36	0.75	–	–	–	–	–
$n = 6$	0.119	0.26	0.442	0.8	–	–	–	–
$n = 7$	0.096	0.205	0.333	0.5	0.833	–	–	–
$n = 8$	0.08	0.169	0.27	0.389	0.544	0.857	–	–
$n = 9$	0.069	0.144	0.228	0.322	0.434	0.578	0.875	–
$n = 10$	0.061	0.126	0.197	0.275	0.364	0.47	0.606	0.889

Table 11.1: Values of $K(C)$ sufficient for the existence of a solution for the given values of n and r.

The second constraint $g(\alpha) > 0$ leads to

$$K(C) < 1 - \frac{(n - r - 1)}{(n - 1)}. \tag{11.19}$$

To decide which constraint of the two (11.18) and (11.19) is more restrictive, let us consider two cases:

(a) $r = n - 2$,

(b) $0 < r \leq n - 3$.

When $r = n - 2$, then $f(\alpha) = \alpha g(\alpha)$, which means that $f(\alpha)$ and $g(\alpha)$ take the 0 value for the same values of the argument α. Thus, both criteria are equivalent. If $0 < r \leq n - 3$, one can prove that (11.18) is more restrictive than (11.19), i.e., whenever (11.18) holds, the inequality (11.19) is also true [112, p. 119–120]. Thus, it is enough to consider the first criterion. The fact that E is *an M-matrix* immediately implies that there is its inversion $E^{-1} \geq 0$. As F is diagonal, it also has its inversion F^{-1}. Thus, in particular, $\overline{C}^{-1} = E^{-1}F^{-1} \geq 0$. Based on the first condition of Theorem 20, we conclude that $\overline{C}$ is *an M-matrix*, and hence $\overline{C}w = b$ has a unique strictly positive solution.   □

The constraint (11.7) formulated in the above Theorem allows the immediate construction of a map containing values of K below which there is a guarantee that (11.3) has a solution (Figure 11.2). For example, if there are seven alternatives where four alternatives are known and three are new, the inconsistency $K(C)$ below 0.5 guarantees the existence of a solution (Table 11.1).

Looking at Figure 11.2 (as well as Table 11.1), it is worth noting that the more reference alternatives (compared to unknown alternatives), the greater the admissible value of inconsistency that guarantees the existence of a solution. This regularity is in line with the observation that with a small number of reference alternatives, even a small inconsistency (small error) can cause a large disturbance of the final results. Conversely, when the number of reference alternatives is relatively large, the possible error in comparing the two alternatives is less important for the final ranking.

It is also worth noting that there are always such n and r for which there is a solution of (11.3) with some non-zero $K(C)$. Formally, for any $r, n \in \mathbb{N}_+$, where $0 < r \leq n - 2$, the right side of (11.18) is greater than 0. To prove it, it is enough to show that

$$\frac{\left(1 + \sqrt{1 + 4(n-1)(n-r-2)}\right)}{2(n-1)} < 1 \tag{11.20}$$

for $n \geq 3$. Indeed, since $\sqrt{1 + 4(n-1)(n-r-2)} \leq \sqrt{1 + 4(n-1)(n-3)}$, the above reduces to

$$\frac{\left(1 + \sqrt{1 + 4(n-1)(n-3)}\right)}{2(n-1)} < 1. \tag{11.21}$$

Hence,

$$\sqrt{1 + 4(n-1)(n-3)} < 2n - 3 \tag{11.22}$$

then

$$4(n-1)(n-3) < (2n-3)^2 - 1, \tag{11.23}$$

i.e.,

$$4(n-1)(n-3) < 4(n-1)(n-2). \tag{11.24}$$

For every $n > 1$ the above inequality boils down to:

$$n - 3 < n - 2. \tag{11.25}$$

As (11.25) is always true, the inequality (11.20) holds for any $n \geq 3$.

Koczkodaj and *Szarek* [103] recommend 1/3 as the acceptance level for Koczkodaj's inconsistency index. Thus, the criterion (11.7) can be written as:

$$\frac{1}{3} < L(n,r). \tag{11.26}$$

It is easy to prove[4] that the above inequality is equivalent to

$$r > h(n), \quad \text{where } h(n) \overset{df}{=} n - 2 - \frac{(4(n-1) - 3)^2 - 9}{36(n-1)} \tag{11.27}$$

---

[4]Consider the series of the following identities: $\frac{1}{3} < 1 - \frac{1 + \sqrt{1 + 4(n-1)(n-r-2)}}{2(n-1)} \Leftrightarrow 1 + \sqrt{1 + 4(n-1)(n-r-2)} < \frac{4(n-1)}{3} \Leftrightarrow 3\sqrt{1 + 4(n-1)(n-r-2)} < 4(n-1) - 3 \Leftrightarrow 9(1 + 4(n-1)(n-r-2)) < (4(n-1) - 3)^2 \Leftrightarrow 36(n-1)(n-r-2) < (4(n-1) - 3)^2 - 9 \Leftrightarrow n - 2 - \frac{(4(n-1)-3)^2 - 9}{36(n-1)} < r$.

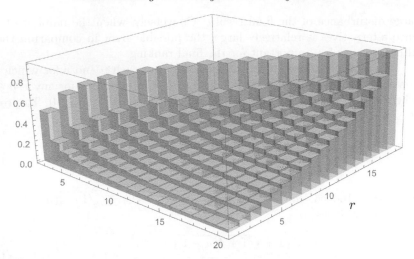

Figure 11.2: Values $K(C)$ below the surface guarantee the existence of the additive AHRE solution

Thus, providing the acceptable inconsistency $K(C) < 1/3$ we may compute the number of reference alternatives guaranteeing the existence of an admissible (real and positive) solution (Table 11.2).

n	3	4	5	6	7	8	9	10	11	12	13	14	15
$r = \lceil h(n) \rceil$	1	2	2	3	3	4	5	5	6	6	7	7	8

Table 11.2: The desired number of reference alternatives providing that $K(C) < 1/3$.

In general, the ratio n and $h(n)$ tends to $5/9$, i.e.,

$$\lim_{n \to \infty} \frac{h(n)}{n} = \frac{5}{9} \approx 0.5556 \qquad (11.28)$$

In other words, for not too high inconsistency $K(C)$, the desired number of reference alternatives is slightly more (exactly 55.56%) than half.

Example 23. *Condition (11.7) is sufficient but not necessary. This means that there are AHRE models in which $K(C) > L(n, r)$ but still the AHRE provides an admissible solution. Let us consider a PC matrix C_2 for an AHRE problem where the reference alternatives a_5, a_6 and a_7 have the following pri-*

orities $w(a_5) = 3, w(a_6) = 4$ *and* $w(a_7) = 5$:

$$
C_2 = \begin{pmatrix}
1. & 0.617 & 3.021 & 1.96 & 1.41 & 2.078 & 0.079 \\
1.62 & 1. & 6.747 & 21.815 & 2.99 & 38.938 & 0.108 \\
0.331 & 0.148 & 1. & 0.188 & 0.268 & 4.459 & 0.052 \\
0.511 & 0.046 & 5.313 & 1. & 1.2 & 3.926 & 0.059 \\
0.71 & 0.334 & 3.726 & 0.837 & 1. & 0.75 & 0.6 \\
0.481 & 0.026 & 0.224 & 0.255 & 1.33 & 1. & 0.8 \\
12.722 & 9.287 & 19.351 & 16.918 & 1.67 & 1.25 & 1.
\end{pmatrix}.
$$

It is easy to compute that the local inconsistency given as $K(C_2)$ *is high. Indeed* $K(C_2) = 0.996543$ *(Saaty* $CI(C_2) = 0.789$*). On the other hand, the unscaled solution of the auxiliary equation*

$$
\begin{pmatrix}
1. & -0.102 & -0.503 & -0.325 \\
-0.269 & 1. & -1.12 & -3.635 \\
-0.055 & -0.024 & 1. & -0.031 \\
-0.085 & -0.007 & -0.885 & 1.
\end{pmatrix}
\begin{pmatrix}
w(a_1) \\
\vdots \\
\vdots \\
w(a_4)
\end{pmatrix}
=
\begin{pmatrix}
2.155 \\
27.543 \\
3.15 \\
3.264
\end{pmatrix}
$$

derived from C_2 *and the reference alternatives is as follows:*

$$
\begin{pmatrix}
w(a_1) \\
\vdots \\
\vdots \\
w(a_4)
\end{pmatrix}
=
\begin{pmatrix}
17.124 \\
79.351 \\
6.4 \\
10.998
\end{pmatrix}.
$$

As the received vector is real and positive, the obtained result is admissible.

Example 24. *When we increase the inconsistency of the model to an even higher level, we may indeed get a model which does not have an admissible solution. For example, let the PC matrix* C_3 *for AHRE be given as:*

$$
C_3 = \begin{pmatrix}
1. & 0.014 & 109.35 & 0.135 & 17.247 & 2.373 & 1.3 \\
73.094 & 1. & 14.011 & 32.741 & 23.775 & 175.057 & 0.098 \\
0.009 & 0.071 & 1. & 0.116 & 4.76 & 7.258 & 0.013 \\
7.392 & 0.03 & 8.578 & 1. & 13.373 & 5.082 & 0.018 \\
0.058 & 0.042 & 0.21 & 0.075 & 1. & 0.75 & 0.6 \\
0.421 & 0.006 & 0.138 & 0.197 & 1.33 & 1. & 0.8 \\
0.766 & 10.25 & 78.211 & 56.713 & 1.67 & 1.25 & 1.
\end{pmatrix}.
$$

In such a case $K(C_3) = 0.9993$ *and Saaty's* $CI(C_3) = 2.709$*. Thus, the inconsistency is even higher than in* C_3*. In such a situation, although the solution exists, it is negative. Indeed the auxiliary equation (11.3) takes the* *form:*

$$
\begin{pmatrix}
1. & -0.002 & -18.225 & -0.022 \\
-12.182 & 1. & -2.335 & -5.456 \\
-0.001 & -0.011 & 1. & -0.019 \\
-1.23 & -0.005 & -1.43 & 1.
\end{pmatrix}
\begin{pmatrix}
w(a_1) \\
\vdots \\
\vdots \\
w(a_4)
\end{pmatrix}
=
\begin{pmatrix}
11.293 \\
128.673 \\
7.23 \\
10.087
\end{pmatrix}.
$$

One can easily compute that the solution is set to

$$
\begin{pmatrix}
w(a_1) \\
\vdots \\
\vdots \\
w(a_4)
\end{pmatrix}
=
\begin{pmatrix}
-47.482 \\
-766.23 \\
-3.059 \\
-56.686
\end{pmatrix}.
$$

As the negative ranking values of the ranking have no interpretation, we have to consider C_3 as too inconsistent to be used in the AHRE model.

Despite the fact that the last example demonstrates that AHRE does not always have an acceptable solution, such cases seem to be rare in practice. The high inconsistency that occurs in such cases disqualifies this type of data as a source of ranking data.

11.4   Geometric HRE

The idea of AHRE comes from two observations. The first one is that the priorities of some alternatives can already be known from the very beginning, whereas the second one is that the priority of a single alternative can be expressed as the arithmetic weighted mean of priorities of other alternatives (11.1). The arithmetic mean underlying the second assumption, however, can be replaced by the weighted geometric mean of priorities. Hence, (11.1) obtains the following non-linear form:

$$
w(a_i) = \left(\prod_{\substack{j=1 \\ i \neq j}}^{n} c_{ij} w(a_j) \right)^{\frac{1}{n-1}}, \tag{11.29}
$$

for $i = 1, \ldots, n$. After raising both sides to the power of $n - 1$ we get a non-linear equation system in the form:

$$w^{n-1}(a_1) = c_{1,2}w(a_2)\cdot\ldots\ldots\cdot c_{1,n}w(a_n)$$
$$w^{n-1}(a_2) = c_{2,1}w(a_1)\cdot c_{2,3}w(a_3)\cdot\ldots\cdot c_{2,n}w(a_n)$$
$$\ldots\ldots\ldots\ldots\ldots\ldots\ldots\ldots$$
$$w^{n-1}(a_k) = c_{k,1}w(a_1)\cdot\ldots\cdot c_{k,n-1}w(a_{n-1})$$

$$(11.30)$$

Similarly as for the AHRE, let $a_{k+1},\ldots,a_n \in A_K$ be reference alternatives. Hence, the values $w(a_{k+1}),\ldots,w(a_n)$ are fixed and known, and as follows $c_{ij}w(a_j)$ are initially known constants. Let us denote:

$$g_i = \prod_{j=k+1}^{n} c_{ij}w(a_j), \qquad (11.31)$$

for $i = 1,\ldots,k$. Thus, the non-linear equation system (11.30) can be written as:

$$w^{n-1}(a_1) = c_{1,2}w(a_2)\cdot\ldots\ldots\cdot c_{1,k}w(a_k)\cdot g_1$$
$$w^{n-1}(a_2) = c_{2,1}w(a_1)\cdot c_{2,3}w(a_3)\cdot\ldots\cdot c_{2,k}w(a_k)\cdot g_2$$
$$\ldots\ldots\ldots\ldots\ldots\ldots\ldots\ldots\ldots$$
$$w^{n-1}(a_k) = c_{k,1}w(a_1)\cdot\ldots\cdot c_{k,k-1}w(a_{k-1})\cdot g_k$$

$$(11.32)$$

Let us replace the above equations by the linear equation system using logarithmic transformation. Hence, let $\log_\xi w(a_i) \stackrel{df}{=} \widehat{w}(a_i)$, $\log_\xi c_{ij} \stackrel{df}{=} \widehat{c}_{ij}$ and $\log_\xi g_j \stackrel{df}{=} \widehat{g}_j$ for some real constant $\xi \in \mathbb{R}_+$. Then (11.32) obtains the form:

$$(n-1)\widehat{w}(a_1) = \widehat{c}_{1,2} + \widehat{w}(a_2) + \ldots + \widehat{c}_{1,k} + \widehat{w}(a_k) + \widehat{g}_1$$
$$(n-1)\widehat{w}(a_2) = \widehat{c}_{2,1} + \widehat{w}(c_1) + \ldots + \widehat{c}_{2,k} + \widehat{w}(a_k) + \widehat{g}_2$$
$$\ldots\ldots\ldots\ldots\ldots\ldots\ldots\ldots\ldots$$
$$(n-1)\widehat{w}(a_k) = \widehat{c}_{k,1} + \widehat{w}(a_1) + \ldots + \widehat{c}_{k,k-1} + \widehat{w}(a_{k-1}) + \widehat{g}_k$$

$$(11.33)$$

The logarithmized entries $\widehat{c}_{i,j}$ of a PC matrix are also constant values. We can therefore group the constant values in each row once again. As a result, we obtain:

$$(n-1)\widehat{w}(a_1) - \sum_{i=2}^{k} \widehat{w}(a_i) = b_1$$
$$(n-1)\widehat{w}(a_2) - \sum_{i=1, i\neq 2}^{k} \widehat{w}(a_i) = b_2$$
$$\ldots\ldots\ldots\ldots\ldots\ldots\ldots$$
$$(n-1)\widehat{w}(a_k) - \sum_{i=1}^{k-1} \widehat{w}(a_i) = b_k$$

$$(11.34)$$

where $b_i \stackrel{df}{=} \sum_{j=1, j\neq i}^{k} \widehat{c}_{i,j} + \widehat{g}_i$ for $i = 1,\ldots,k$. The above linear equation system can be written down in the form of the following matrix equation:

$$\widehat{C}\widehat{w} = b, \qquad (11.35)$$

where

$$
\widehat{C} = \begin{bmatrix} (n-1) & -1 & \cdots & & -1 \\ \vdots & \ddots & & & \vdots \\ \vdots & & \ddots & & \vdots \\ -1 & -1 & \cdots & & (n-1) \end{bmatrix}, \tag{11.36}
$$

$$
\widehat{w} = \begin{bmatrix} \widehat{w}(a_1) \\ \widehat{w}(a_2) \\ \vdots \\ \widehat{w}(a_k) \end{bmatrix}, \text{ and } b = \begin{bmatrix} b_1 \\ b_2 \\ \vdots \\ b_k \end{bmatrix}. \tag{11.37}
$$

Solving the above equation simultaneously, we get the solution of the original nonlinear equation (11.30). Indeed, when computed, the original priority vector is given as:

$$
w = \begin{pmatrix} \xi^{\widehat{w}(a_1)} \\ \vdots \\ \vdots \\ \xi^{\widehat{w}(a_k)} \end{pmatrix}.
$$

Unlike in AHRE, there is always an acceptable solution of (11.35). Hence, the geometric HRE (GHRE) always provides the user with an appropriate ranking of alternatives.

Example 25. *GHRE is resistant to inconsistency. Because there is always a solution in the GHRE method, it can be used even if AHRE fails. Let us consider the matrix M_3 (Example 24). The auxiliary linear matrix equation (11.35) takes the form*

$$
\begin{pmatrix} 6. & -1. & -1. & -1. \\ -1. & 6. & -1. & -1. \\ -1. & -1. & 6. & -1. \\ -1. & -1. & -1. & 6. \end{pmatrix} \begin{pmatrix} \widehat{w}(a_1) \\ \vdots \\ \vdots \\ \widehat{w}(a_4) \end{pmatrix} = \begin{pmatrix} 1.7 \\ 5.254 \\ -1.47 \\ 0.316 \end{pmatrix},
$$

where the logarithm base is $\xi = 10$. Since

$$
\begin{pmatrix} 6. & -1. & -1. & -1. \\ -1. & 6. & -1. & -1. \\ -1. & -1. & 6. & -1. \\ -1. & -1. & -1. & 6. \end{pmatrix}^{-1} = \begin{pmatrix} 0.19 & 0.048 & 0.048 & 0.048 \\ 0.048 & 0.19 & 0.048 & 0.048 \\ 0.048 & 0.048 & 0.19 & 0.048 \\ 0.048 & 0.048 & 0.048 & 0.19 \end{pmatrix},
$$

then

$$
\begin{pmatrix} \widehat{w}(a_1) \\ \vdots \\ \vdots \\ \widehat{w}(a_4) \end{pmatrix} = \begin{pmatrix} 0.19 & 0.048 & 0.048 & 0.048 \\ 0.048 & 0.19 & 0.048 & 0.048 \\ 0.048 & 0.048 & 0.19 & 0.048 \\ 0.048 & 0.048 & 0.048 & 0.19 \end{pmatrix} \begin{pmatrix} 1.7 \\ 5.254 \\ -1.47 \\ 0.316 \end{pmatrix}.
$$

i.e.,

$$
\begin{pmatrix} \widehat{w}(a_1) \\ \vdots \\ \vdots \\ \widehat{w}(a_4) \end{pmatrix} = \begin{pmatrix} 0.519 \\ 1.03 \\ 0.067 \\ 0.322 \end{pmatrix}.
$$

Thus, the final ranking is

$$
\begin{pmatrix} w(a_1) \\ \vdots \\ \vdots \\ w(a_4) \\ w(a_5) \\ w(a_6) \\ w(a_7) \end{pmatrix} = \begin{pmatrix} \xi^{\widehat{w}(a_1)} \\ \vdots \\ \vdots \\ \xi^{\widehat{w}(a_4)} \\ 3 \\ 4 \\ 5 \end{pmatrix} = \begin{pmatrix} 3.308 \\ 10.64 \\ 1.17 \\ 2.098 \\ 3 \\ 4 \\ 5 \end{pmatrix}.
$$

Thus, taking into account that $w(a_5) = 3, w(a_6) = 4$ and $w(a_7) = 5$ the total order of alternatives is as follows: $a_2, a_7, a_6, a_1, a_5, a_4$ and a_3. Of course, due to the high inconsistency, the result can be disputable. The next few examples will show more practical applications of the HRE method.

11.5   Existence of a solution for geometric HRE

At first glance, one can see that $\widehat{C}$ has a special form. Indeed, only elements on a diagonal are positive, while all the other elements are negative. Moreover, diagonal elements are much greater in terms of absolute value than the elements outside the diagonal. Therefore, all we have to do is to check whether $\widehat{C}$ (11.36) has a strongly dominant diagonal, i.e.,

$$
|\widehat{c}_{ii}| > \sum_{\substack{j=1 \\ i \neq j}}^{k} |c_{ij}|,
$$

for $i = 1, \ldots, k$. Since $\widehat{c}_{ij} = n - 1$ and $c_{ij} = -1$ (for $i \neq j$) then the above constraint obtains the form

$$n - 1 > \sum_{\substack{j=1 \\ i \neq j}}^{k} 1 = k - 1.$$

Of course, it is always true as $n > k$. This means that $\widehat{C}$ has a strictly positive diagonal. Since it is also symmetric, then it is a positive definite matrix [146, p. 29] and as such is nonsingular [146, p. 28]. Hence, there exists $\widehat{C}^{-1} > 0$, which implies that $\widehat{w} = \widehat{C}^{-1}b$. Because both $\widehat{C}^{-1}$ and $b > 0$ then also $\widehat{w} > 0$. Thus (11.35) has a unique, positive solution. It is worth noting that due to Theorem 20, $\widehat{C}$ is also an M matrix.

The matrix $\widehat{C}^{-1} = [\check{c}_{ij}]$ also has a very specific form. Indeed

$$\check{c}_{ij} = \begin{cases} \frac{n^{k-1} - (k-1)n^{k-2}}{n^k - kn^{k-1}} & i = j \\ \frac{n^{k-2}}{n^k - kn^{k-1}} & j \neq j \end{cases}.$$

This implies that the exact solution for (11.35) is given as:

$$\widehat{w}(a_i) = \frac{n^{k-1} - (k-1)n^{k-2}}{n^k - kn^{k-1}} \left(\sum_{\substack{j=1 \\ ,j \neq i}}^{k} \widehat{c}_{i,j} + \widehat{g}_i \right) +$$

$$+ \frac{n^{k-2}}{n^k - kn^{k-1}} \sum_{\substack{r=1 \\ r \neq i}}^{k} \left(\sum_{\substack{j=1 \\ j \neq r}}^{k} \widehat{c}_{r,j} + \widehat{g}_r \right),$$

for $i = 1, \ldots, k$. Hence, despite the need to solve a linear equation system, calculating the result in GHRE is easy ($w(a_i) = \xi^{\widehat{w}(a_i)}$ for $i = 1, \ldots, k$) and does not require specialized software.

11.6   Is geometric HRE optimal?

GHRE is optimal just like GMM. This means that it minimizes the logarithmic errors providing, of course, that the selected priorities are a priori known. The reasoning allowing the optimality of GHRE to be shown is quite similar to the one used for GMM (Chapter 3.2.3). For this reason, in the following discussion, some simple transformations were omitted in the hope that an attentive reader will find them in Chapter 3.2.

As in GHRE, there are k unknown priorities, hence the error function will only have k variables. Let $\widehat{\mathcal{E}}$be defined as

$$\widehat{\mathcal{E}}(u_1, \ldots, u_k) \overset{df}{=} \sum_{i=1}^{n} \sum_{j=1}^{n} e_{ij}$$

where

$$e_{ij} = (\ln c_{ij} - u_i + u_j)^2,$$

$u_i = \log w(a_i)$ and the values $u_{k+1}, \ldots, u_n$ are known and treated as constants. The demand for optimality of GHRE can be formulated as a theorem.

Theorem 22. *The GHRE priority vector minimizes $\widehat{\mathcal{E}}$.*

Proof. According to the first order condition for the existing minimum of the given function in x, its first derivative needs to be 0 in x. In the case of $\widehat{\mathcal{E}}$ we demand that

$$\begin{pmatrix} \frac{\partial \widehat{\mathcal{E}}}{\partial u_1} \\ \frac{\partial \widehat{\mathcal{E}}}{\partial u_2} \\ \vdots \\ \frac{\partial \widehat{\mathcal{E}}}{\partial u_k} \end{pmatrix} = 0. \tag{11.38}$$

Hence

$$\frac{\partial \widehat{\mathcal{E}}}{\partial u_i} = \frac{\partial}{\partial u_i} \sum_{i=1}^{n} \sum_{j=1}^{n} (\ln c_{ij} - u_i + u_j)^2,$$

for $i = 1, \ldots, k$. By solving the system of k equations in the form:

$$\frac{\partial}{\partial u_i} \sum_{k=1}^{n} \sum_{j=1}^{n} (\ln c_{ij} - u_k + u_j)^2 = 0, \tag{11.39}$$

After calculating the partial derivatives, we get the system of k equations in the form:

$$\sum_{\substack{j=1 \\ i \neq j}}^{n} 2 \left(\ln \frac{c_{j,i}}{c_{i,j}} \right) + 4(n-1)u_i - \sum_{\substack{j=1 \\ j \neq i}}^{n} 4u_j = 0.$$

We may transform it to

$$\sum_{\substack{j=1 \\ i \neq j}}^{n} 4 \left(\ln \frac{1}{c_{ij}} \right) + 4(n-1)u_i - \sum_{\substack{j=1 \\ i \neq j}}^{n} 4u_j = 0,$$

for $i = 1, \ldots, k$. The above equation (Theorem 1) is equivalent to:

$$nu_i = \sum_{j=1}^{n} u_j - \sum_{j=1}^{n} \ln \frac{1}{c_{ij}}.$$

Hence,

$$n \ln w(a_i) = \sum_{j=1}^{n} \ln c_{ij} w(a_j).$$

By subtracting $\ln w(a_i)$ we get

$$(n-1) \ln w(a_i) = \sum_{\substack{j=1 \\ i \neq j}}^{n} \ln c_{ij} w(a_j),$$

i.e.,

$$\ln w(a_i) = \frac{1}{n-1} \sum_{\substack{j=1 \\ i \neq j}}^{n} \ln c_{ij} w(a_j).$$

$$\ln w(a_i) = \frac{1}{n-1} \ln \prod_{\substack{j=1 \\ i \neq j}}^{n} c_{ij} w(a_j),$$

and finally

$$w(a_i) = \left(\prod_{\substack{j=1 \\ i \neq j}}^{n} c_{ij} w(a_j) \right)^{\frac{1}{n-1}},$$

for $i = 1, \ldots, k$. This means (see 11.29) that, at the point being the solution of the GHRE method, the function $\widehat{\mathcal{E}}$ reaches its extreme. We just have to confirm that this is really the minimum. For this purpose, let us consider the Hessian matrix $H_{\widehat{\mathcal{E}}}$ of $\widehat{\mathcal{E}}$. We therefore have:

$$H_{\widehat{\mathcal{E}}} = \begin{pmatrix} \frac{\partial^2 \widehat{\mathcal{E}}}{\partial^2 x_1} & \frac{\partial^2 \widehat{\mathcal{E}}}{\partial x_1 \partial x_2} & \cdots & \frac{\partial^2 \widehat{\mathcal{E}}}{\partial x_1 \partial x_k} \\ \frac{\partial^2 \widehat{\mathcal{E}}}{\partial x_2 \partial x_1} & \frac{\partial^2 \widehat{\mathcal{E}}}{\partial^2 x_2} & \cdots & \vdots \\ \vdots & \cdots & \ddots & \vdots \\ \frac{\partial^2 \widehat{\mathcal{E}}}{\partial x_k \partial x_1} & \cdots & \cdots & \frac{\partial^2 \widehat{\mathcal{E}}}{\partial^2 x_k} \end{pmatrix},$$

which immediately reduces to

$$H_{\widehat{\mathcal{E}}} = \begin{pmatrix} 4(n-1) & -4 & \cdots & -4 \\ -4 & 4(n-1) & \cdots & \vdots \\ \vdots & \cdots & \ddots & \vdots \\ -4 & \cdots & \cdots & 4(n-1) \end{pmatrix}.$$

Since the dimensions of $H_{\widehat{\mathcal{E}}}$ are $k \times k$, the analysis of $H_{\widehat{\mathcal{E}}}$ is even more straightforward than in the case of GMM (Theorem 1). For the purposes of this reasoning, it is sufficient to note that $H_{\widehat{\mathcal{E}}} = [h_{ij}]$ is strictly diagonally dominant, i.e.,

$$|h_{ii}| > \sum_{\substack{j=1 \\ j \neq i}}^{k} |h_{ij}| .$$

which is true as

$$|h_{ii}| = 4(n-1), \quad \sum_{\substack{j=1 \\ j \neq i}}^{k} |h_{ij}| = 4(k-1), \quad \text{and} \quad n > k.$$

Thus, $H_{\widehat{\mathcal{E}}}$ is a positive definite matrix [146, p. 29]. Thus, $\widehat{\mathcal{E}}$ has a strict local minimum [190, p. 98] in $w = [w(a_1), \ldots, w(a_k)]^T$ such that $w(a_i) = \left(\prod_{\substack{j=1 \\ i \neq j}}^{n} c_{ij} w(a_j) \right)^{1/n-1}$. Thus, GHRE minimizes $\widehat{\mathcal{E}}(y_1, \ldots, y_k)$. □

It is worth noting that geometric HRE ensures optimal priority values only for the unknown alternatives $a_1, \ldots, a_k \in A_U$. The priorities of known (reference) alternatives remain unchanged, thus, they cannot be optimal in the above, logarithmic least squares sense. This "partial" optimality is the price we have to pay for leaving the previously prepared ranking unchanged. However, thanks to this approach, we gain a method resistant to the rank reversal problem, i.e., the situations in which adding new (often the same as already existing) alternatives changes the previously created ranking.

11.7 Illustrative examples

Example 26. *A certain cable television network wants to launch a new TV series. It considers purchasing one new series out of five different ones $a_1, \ldots, a_5$ produced in the United States. So far it has broadcast three similar series. Through market research, the approximate size of their audience is known. They are $271,000$, $278,000$ and $233,000$ viewers out of 1.5 million subscribers for the series a_6, a_7 and a_8 respectively. The prices of all serial productions are comparable. In order to select possibly the most profitable TV series, the television network asks a few media experts. During the meeting they prepare a PC matrix C_s, representing the relationship between the attractiveness of all the considered series.*

$$C_s = \begin{pmatrix} 1 & \frac{23}{24} & \frac{25}{11} & 2 & \frac{21}{20} & 3 & \frac{5}{3} & \frac{7}{2} \\ \frac{24}{23} & 1 & \frac{23}{8} & \frac{23}{10} & \frac{6}{5} & \frac{9}{2} & \frac{22}{13} & \frac{13}{4} \\ \frac{11}{25} & \frac{8}{23} & 1 & \frac{17}{21} & \frac{11}{25} & \frac{17}{14} & \frac{12}{25} & \frac{16}{13} \\ \frac{1}{2} & \frac{10}{23} & \frac{21}{17} & 1 & \frac{10}{19} & \frac{25}{17} & \frac{11}{15} & \frac{7}{6} \\ \frac{20}{21} & \frac{5}{6} & \frac{25}{11} & \frac{19}{10} & 1 & \frac{16}{5} & \frac{8}{7} & \frac{13}{5} \\ \frac{1}{3} & \frac{2}{9} & \frac{14}{17} & \frac{17}{25} & \frac{5}{16} & 1 & \mathbf{\frac{271}{278}} & \mathbf{\frac{271}{233}} \\ \frac{3}{5} & \frac{13}{22} & \frac{25}{12} & \frac{15}{11} & \frac{7}{8} & \mathbf{\frac{278}{271}} & 1 & \mathbf{\frac{278}{233}} \\ \frac{2}{7} & \frac{4}{13} & \frac{13}{16} & \frac{6}{7} & \frac{5}{13} & \mathbf{\frac{233}{271}} & \mathbf{\frac{233}{278}} & 1 \end{pmatrix}.$$

As the values of attractiveness for a_6, a_7 and a_8 are known (they are approximated by the number of people watching the given TV series), thus the appropriate comparisons c_{ij} (bold) for $i, j = 6, 7, 8$ are a priori known and are not the subject of expert assessment. For example:

$$c_{6,7} = \frac{271000}{278000} = \frac{271}{278} = 0.974 \ .$$

The other entries of C_s are the result of expert considerations. To find a solution with the help of additive HRE, the following system of linear equations must be solved:

$$\begin{pmatrix} 1. & -0.136 & -0.324 & -0.285 & -0.15 \\ -0.149 & 1. & -0.41 & -0.328 & -0.171 \\ -0.062 & -0.049 & 1. & -0.115 & -0.062 \\ -0.071 & -0.062 & -0.176 & 1. & -0.075 \\ -0.136 & -0.119 & -0.324 & -0.271 & 1. \end{pmatrix} \begin{pmatrix} w(a_1) \\ w(a_2) \\ w(a_3) \\ w(a_4) \\ w(a_5) \end{pmatrix} = \begin{pmatrix} 298833 \\ 349602 \\ 107040 \\ 124890 \\ 255816 \end{pmatrix} \ .$$

Solving the system by calculating the inverse matrix, we obtain:

$$\begin{pmatrix} w(a_1) \\ w(a_2) \\ w(a_3) \\ w(a_4) \\ w(a_5) \end{pmatrix} = \begin{pmatrix} 1.17 & 0.267 & 0.692 & 0.585 & 0.308 \\ 0.32 & 1.17 & 0.814 & 0.664 & 0.349 \\ 0.123 & 0.105 & 1.17 & 0.239 & 0.128 \\ 0.146 & 0.128 & 0.356 & 1.17 & 0.154 \\ 0.276 & 0.244 & 0.666 & 0.553 & 1.17 \end{pmatrix} \begin{pmatrix} 298833 \\ 349602 \\ 107040 \\ 124890 \\ 255816 \end{pmatrix}$$

and finally,

$$\begin{pmatrix} w(a_1) \\ w(a_2) \\ w(a_3) \\ w(a_4) \\ w(a_5) \end{pmatrix} = \begin{pmatrix} 668179 \\ 762937 \\ 261151 \\ 311725 \\ 606951 \end{pmatrix} \ .$$

As a result of the calculations made, a_2 was chosen as the most promising TV series with the expected audience of about 762,937 subscribers.

Example 27. *An appraiser was asked to estimate the price of two paintings a_1 and a_2 painted by a well-known contemporary artist. The expert decided*

to make an assessment based on the transaction prices of five other paintings
$a_3, \ldots, a_7$ *similar to the rated canvases in painting school, condition and size.*
The prices of the reference paintings were $w(a_3) = 431\,753$ *,* $w(a_4) = 54\,236$*,*
$w(a_5) = 52\,722$*,* $w(a_6) = 122\,991$*, and* $w(a_7) = 163\,253$*. To this end, he/she*
compared the relative value of all paintings and prepared the following PC
matrix:

$$
C_p = \begin{pmatrix}
1. & 1.49 & 0.104 & 0.706 & 0.974 & 0.656 & 0.482 \\
0.672 & 1. & 0.065 & 0.493 & 0.33 & 0.141 & 0.179 \\
9.597 & 15.282 & 1. & 7.961 & 8.189 & 3.51 & 2.645 \\
1.42 & 2.026 & 0.126 & 1. & 1.03 & 0.441 & 0.332 \\
1.03 & 3.026 & 0.122 & 0.972 & 1. & 0.429 & 0.323 \\
1.52 & 7.101 & 0.285 & 2.268 & 2.333 & 1. & 0.753 \\
2.074 & 5.591 & 0.378 & 3.01 & 3.096 & 1.33 & 1.
\end{pmatrix}.
$$

The reference paintings whose prices were previously known were not compared
with each other. The values of their comparisons were calculated on the basis
of their prices, i.e.,

$$
c_{3,4} = \frac{w(a_3)}{w(a_4)} = \frac{431\,753}{54\,236} = 7.961
$$

or

$$
c_{4,5} = \frac{w(a_4)}{w(a_6)} = \frac{54\,236}{122\,991} = 0.441,
$$

and so on. The HRE matrix equation (11.3) is as follows:

$$
\begin{pmatrix} 1 & -0.247861 \\ -0.11207 & 1 \end{pmatrix} \begin{pmatrix} w(a_1) \\ w(a_2) \end{pmatrix} = \begin{pmatrix} 49011.6 \\ 19825.9 \end{pmatrix}.
$$

Thus, the estimated values of paintings a_1 *and* a_2 *are*

$$
\begin{pmatrix} w(a_1) \\ w(a_2) \end{pmatrix} = \begin{pmatrix} 55\,466.5 \\ 26\,042. \end{pmatrix}.
$$

Example 28. *Eight candidates* $a_1, \ldots, a_8$ *applied for a position in the com-*
pany, but four of them were rejected for formal reasons. Those whose ap-
plications were rejected were a_2, a_4, a_7, a_8*. Shortly after the expert meeting*
during which the winner was selected using the pairwise comparison method,
some unsuccessful candidates completed their documentation and resubmit-
ted their applications. As a result of resubmission, three candidates a_2, a_4
and a_7 *were allowed to participate in the recruitment process. Initially, the*
recruitment process was attended by a_1, a_3, a_5 *and* a_6*. As a result of the*
decision procedure, the candidates received the following numerical scores:
$w(a_1) = 31, w(a_3) = 21, w(a_5) = 33, w(a_6) = 27$*. After extending the list*
of candidates by another three participants, the experts decided to compare
them with the previous four candidates. Experts chose the GHRE method to

*extend the ranking. As a result of the pairwise comparison of four candidates
already assessed and three added later, the following matrix was obtained:*

$$
\begin{pmatrix}
1 & \frac{1}{2} & \mathbf{\frac{31}{21}} & \frac{4}{13} & \mathbf{\frac{31}{33}} & \mathbf{\frac{31}{27}} & \frac{1}{2} \\
2 & 1 & \frac{1}{2} & \frac{4}{7} & \frac{7}{3} & \frac{5}{9} & 1 \\
\mathbf{\frac{21}{31}} & 2 & 1 & \frac{9}{7} & \frac{7}{11} & \frac{7}{9} & \frac{7}{4} \\
\frac{13}{4} & \frac{7}{4} & \frac{7}{9} & 1 & \frac{10}{3} & 1 & \frac{4}{3} \\
\mathbf{\frac{33}{31}} & \frac{3}{7} & \frac{11}{7} & \frac{3}{10} & 1 & \mathbf{\frac{11}{9}} & \frac{5}{9} \\
\mathbf{\frac{27}{31}} & \frac{9}{5} & \frac{9}{7} & 1 & \frac{9}{11} & 1 & \frac{4}{3} \\
2 & 1 & \frac{4}{7} & \frac{3}{4} & \frac{9}{5} & \frac{3}{4} & 1
\end{pmatrix}.
$$

*Comparisons for a_1, a_3, a_5 and a_6 were prepared as ratios of their previously
established weights (bolded).*

The auxiliary logarithmized matrix equation is

$$
\begin{pmatrix}
6. & -1. & -1. \\
-1. & 6. & -1. \\
-1. & -1. & 6.
\end{pmatrix}
\begin{pmatrix}
\widehat{w}(a_2) \\
\widehat{w}(a_4) \\
\widehat{w}(a_7)
\end{pmatrix}
=
\begin{pmatrix}
5.63312 \\
7.05705 \\
5.82685
\end{pmatrix}.
$$

It leads to the solution

$$
\begin{pmatrix}
\widehat{w}(a_2) \\
\widehat{w}(a_4) \\
\widehat{w}(a_7)
\end{pmatrix}
=
\begin{pmatrix}
1.46605 \\
1.66947 \\
1.49373
\end{pmatrix}.
$$

Thus,

$$
\begin{pmatrix}
w(a_2) \\
w(a_4) \\
w(a_7)
\end{pmatrix}
=
\begin{pmatrix}
10^{\widehat{w}(a_2)} \\
10^{\widehat{w}(a_4)} \\
10^{\widehat{w}(a_7)}
\end{pmatrix}
=
\begin{pmatrix}
29.2452 \\
46.7168 \\
31.1694
\end{pmatrix}.
$$

As a result of the above procedure, the following order of candidates was obtained: $w(a_4) = 46.7$, $w(a_5) = 33$, $w(a_7) = 31.16$, $w(a_1) = 31$, $w(a_2) = 29.25$,
$w(a_6) = 27$, $w(a_3) = 21$. It is worth noting that the order and assessment
of candidates once evaluated did not change as the result of the GHRE procedure. They are not re-evaluated and cannot undermine the validity of the
recalculated result. On the other hand, the other three people a_2, a_4 and a_7
were evaluated by using the pairwise comparison method, in principle, the
same method as the initial four.

Bibliography

[1] K. Ábele-Nagy. *Páros összehasonlítás mátrixok a többszempontú döntéselméletben.* PhD thesis, Budapesti Corvinus Egyetem, October 2019.

[2] J. Aczél and T. L. Saaty. Procedures for synthesizing ratio judgements. *Journal of Mathematical Psychology,* 27(1):93–102, March 1983.

[3] J. Aguarón and J. M. Moreno-Jiménez. The geometric consistency index: Approximated thresholds. *European Journal of Operational Research,* 147(1):137 – 145, 2003.

[4] K. Ahsan and S. Rahman. Green public procurement implementation challenges in Australian public healthcare sector. *Journal of Cleaner Production,* 152:181–197, May 2017.

[5] G. G. Alway. The Distribution of the Number of Circular Triads in Paired Comparisons. *Biometrika,* 49(1/2):265, 1962.

[6] S. Apak, G. G Göğüş, and İ. S. Karakadılar. An Analytic Hierarchy Process Approach with a Novel Framework for Luxury Car Selection. *Procedia - Social and Behavioral Sciences,* 58:1301–1308, October 2012.

[7] B. Aupetit and Ch. Genest. On some useful properties of the Perron eigenvalue of a positive reciprocal matrix in the context of the analytic hierarchy process. *European Journal of Operational Research,* pages 1–6, July 1993.

[8] L. D'Apuzzo B. Cavallo. A general unified framework for pairwise comparison matrices in multicriteria methods. *International Journal of Intelligent Systems,* 24(4):377–398, 2009.

[9] S. M. Baas and H. Kwakernaak. Rating and ranking of multiple-aspect alternatives using fuzzy sets. *Automatica,* 13(1):47–58, January 1977.

[10] J. F. Baldwin and N. C. F. Guild. Comparison of fuzzy sets on the same decision space. *Fuzzy Sets and Systems,* 2(3):213–231, July 1979.

[11] C. A. Bana e Costa and J. Vansnick. A critical analysis of the eigenvalue method used to derive priorities in AHP. *European Journal of Operational Research,* 187(3):1422–1428, June 2008.

[12] C.A. Bana e Costa and J. C.. Vansnick. Macbeth - an interactive path towards the construction of cardinal value fonctions. *International transactions in operational Research*, 1:489–500, 1994.

[13] J. Barzilai. Deriving weights from pairwise comparison matrices. *The Journal of the Operational Research Society*, 48(12):1226–1232, December 1997.

[14] J. Barzilai. Consistency measures for pairwise comparison matrices. *Journal of Multi-Criteria Decision Analysis*, 7(3):123–132, 1998.

[15] J. Barzilai. Preference Function Modelling: The Mathematical Foundations of Decision Theory. In *Trends in Multiple Criteria Decision Analysis*, pages 57–86. Springer US, Boston, MA, January 2010.

[16] A. Berman. *m Applications of M-Matrices*, pages 115–126. Springer US, Boston, MA, 1992.

[17] M. Bernasconi, C. Choirat, and R. Seri. The Analytic Hierarchy Process and the Theory of Measurement. *Management Science*, 56(4):699–711, 2010.

[18] B. Blagojevic, B. Srdjevic, Z. Srdjevic, and T. Zoranovic. Heuristic aggregation of individual judgments in AHP group decision making using simulated annealing algorithm. *Information Sciences*, 330(C):260–273, February 2016.

[19] G. Bortolan and R. Degani. A review of some methods for ranking fuzzy subsets. *Fuzzy Sets and Systems*, 15(1):1–19, February 1985.

[20] S. P. Boyd and L. Vandenberghe. *Convex Optimization*. Cambridge University Press, March 2004.

[21] S. Bozóki. Solution of the least squares method problem of pairwise comparison matrices. *Central European Journal of Operations Research*, 16(4):345–358, 2008.

[22] S. Bozóki. Inefficient weights from pairwise comparison matrices with arbitrarily small inconsistency. *Optimization*, 0(0):1–9, 2014.

[23] S. Bozóki, J. Fülöp, and L. Rónyai. On optimal completion of incomplete pairwise comparison matrices. *Mathematical and Computer Modelling*, 52(1–2):318 – 333, 2010.

[24] S. Bozóki and T. Rapcsák. On Saaty's and Koczkodaj's inconsistencies of pairwise comparison matrices. *Journal of Global Optimization*, 42(2):157–175, 2008.

[25] S. Bozóki and V. Tsyganok. The (logarithmic) least squares optimality of the arithmetic (geometric) mean of weight vectors calculated from all spanning trees for incomplete additive (multiplicative) pairwise comparison matrices. *International Journal of General Systems*, 48(4):362–381, 2019.

[26] I. N. Bronstein, K. A. Semendjajew, G. Musiol, and H. Mühlig. *Handbook of Mathematics*. Springer Verlag, Frankfurt am Main, 5 edition, 2005.

[27] M. Brunelli. *Introduction to the Analytic Hierarchy Process*. Springer-Briefs in Operations Research. Springer International Publishing, Cham, 2015.

[28] M. Brunelli. On the conjoint estimation of inconsistency and intransitivity of pairwise comparisons. *Operations Research Letters*, 44(5):672–675, September 2016.

[29] M. Brunelli. Recent Advances on Inconsistency Indices for Pairwise Comparisons - A Commentary. *Fundam. Inform.*, 144(3-4):321–332, 2016.

[30] M. Brunelli. Studying a set of properties of inconsistency indices for pairwise comparisons. *Annals of Operations Research*, 248(1-2):143–161, March 2016.

[31] M. Brunelli. A survey of inconsistency indices for pairwise comparisons. *International Journal of General Systems*, 47(8):751–771, September 2018.

[32] M. Brunelli, L. Canal, and M. Fedrizzi. Inconsistency indices for pairwise comparison matrices: a numerical study. *Annals of Operations Research*, 211:493–509, February 2013.

[33] M. Brunelli and M. Fedrizzi. Axiomatic properties of inconsistency indices for pairwise comparisons. *Journal of the Operational Research Society*, 66(1):1–15, Jan 2015.

[34] M. Brunelli and M. Fedrizzi. Boundary properties of the inconsistency of pairwise comparisons in group decisions. *European Journal of Operational Research*, 240(3):765 – 773, 2015.

[35] N. Bryson. Group decision-making and the analytic hierarchy process: Exploring the consensus-relevant information content. *Computers & operations research*, 23(1):27–35, January 1996.

[36] Dae-Ho Byun. The AHP approach for selecting an automobile purchase model. *Information & Management*, 38(5):289–297, April 2001.

[37] A. Cayley. A theorem on trees. *Quart. J. Pure Appl. Math.*, 23:376–378, 1889.

[38] V. Čerňanová, W.W. Koczkodaj, and J. Szybowski. Inconsistency of special cases of pairwise comparisons matrices. *International Journal of Approximate Reasoning*, 95:36 – 45, 2018.

[39] Da-Yong Chang. Applications of the extent analysis method on fuzzy AHP. *Elsevier*, 95(3):649–655, December 1996.

[40] E. U. Choo and W. C. Wedley. A common framework for deriving preference values from pairwise comparison matrices. *Computers and Operations Research*, 31(6):893 – 908, 2004.

[41] J. M. Colomer. Ramon Llull: from 'Ars electionis' to social choice theory. *Social Choice and Welfare*, 40(2):317–328, October 2011.

[42] M. Condorcet. Essay on the Application of Analysis to the Probability of Majority Decisions. Paris: Imprimerie Royale, 1785.

[43] A. H. Copeland. A "reasonable" social welfare function. Seminar on applications of mathematics to social sciences, 1951.

[44] G. Crawford and C. Williams. The Analysis of Subjective Judgment Matrices. Technical report, The Rand Corporation, 1985.

[45] G. B. Crawford. The geometric mean procedure for estimating the scale of a judgement matrix. *Mathematical Modelling*, 9(3–5):327 – 334, 1987.

[46] R. Crawford and C. Williams. A note on the analysis of subjective judgement matrices. *Journal of Mathematical Psychology*, 29:387 – 405, 1985.

[47] L. Csató. Axiomatizations of inconsistency indices for triads. *Annals of Operations Research*, 280(1):99–110, September 2019.

[48] H. A. David. *The method of paired comparisons*. A Charles Griffin Book, 1969.

[49] Reinhard Diestel. *Graph theory*. Springer Verlag, 2005.

[50] F. J. Dodd and H. A. Donegan. Comparison of prioritization techniques using interhierarchy mappings. *Journal of the Operational Research Society*, 46(4):492–498, 1995.

[51] Y. Dong, Y. Xu, H. Li, and M. Dai. A comparative study of the numerical scales and the prioritization methods in AHP. *European Journal of Operational Research*, 186(1):229–242, March 2008.

[52] Y. Dong, G. Zhang, W. Hong, and Y. Xu. Consensus models for AHP group decision making under row geometric mean prioritization method. *Decision Support Systems*, 49(3):281–289, June 2010.

[53] Z. Duszak and W. W. Koczkodaj. Generalization of a new definition of consistency for pairwise comparisons. *Information Processing Letters*, 52(5):273 – 276, 1994.

[54] J. S. Dyer. A clarification of remarks on the analytic hierarchy process. *Management Science*, 36(3):274–275, March 1990.

[55] J. S. Dyer. Remarks on the analytic hierarchy process. *Management Science*, 36(3):249–258, 1990.

[56] R. F. Dyer and E. H. Forman. Group decision support with the Analytic Hierarchy Process. *Decision Support Systems*, 8(2):99–124, 1992.

[57] J. R. Emshoff and T. L. Saaty. Applications of the analytic hierarchy process to long range planning processes. *European Journal of Operational Research*, 10(2):131–143, 1982.

[58] M. T. Escobar and J. M. Moreno-Jiménez. Aggregation of Individual Preference Structures in AHP-Group Decision Making. *Group Decision and Negotiation*, 16(4):287–301, July 2007.

[59] A. Farkas, P. Lancaster, and P. Rózsa. Consistency adjustments for pairwise comparison matrices. *Numerical Linear Algebra with Applications*, 10(8):689–700, November 2002.

[60] A. M. Feldman and R. Serrano. *Welfare Economics and Social Choice Theory.* Springer, 2006.

[61] J. Figueira, M. Ehrgott, and S. Greco, editors. *Multiple Criteria Decision Analysis: State of the Art Surveys.* Springer, 2005.

[62] J. S. Finan and W. J. Hurley. Transitive calibration of the AHP verbal scale. *European Journal of Operational Research*, 112(2):367–372, 1999.

[63] E. Forman and K. Peniwati. Aggregating individual judgments and priorities with the analytic hierarchy process. *European Journal of Operational Research*, 108(1):165–169, July 1998.

[64] E. H. Forman. Facts and fictions about the analytic hierarchy process. *Mathematical and Computer Modelling*, 17(4-5):19–26, February 1993.

[65] J. Franek and A. Kresta. Judgment Scales and Consistency Measure in AHP. *Procedia Economics and Finance*, 12:164–173, 2014.

[66] J. Fülöp. *A method for approximating pairwise comparison matrices by consistent matrices.* Springer, 2008.

[67] F. R. Gantmaher. *The theory of matrices.* American Mathematical Society, 2000.

[68] S. I. Gass. Tournaments, Transitivity and Pairwise Comparison Matrices. *The Journal of the Operational Research Society*, 49(6):616–624, June 1998.

[69] S. I. Gass and T. Rapcsák. Singular value decomposition in AHP. *European Journal of Operational Research*, 154(3):573–584, May 2004.

[70] M. Gavalec, J. Ramik, and K. Zimmermann. *Decision Making and Optimization: Special Matrices and Their Applications in Economics and Management.* September 2014.

[71] C. Genest, F. Lapointe, and S. W. Drury. On a Proposal of Jensen for the Analysis of Ordinal Pairwise Preferences Using Saaty's Eigenvector Scaling Method. *Journal of Mathematical Psychology*, 37(4):575–610, 1993.

[72] B. Golany and M. Kress. A multicriteria evaluation of methods for obtaining weights from ratio-scale matrices. *European Journal of Operational Research*, 69(2):210–220, 1993.

[73] B. L. Golden and Q. Wang. *An Alternate Measure of Consistency*, pages 68–81. Springer Berlin Heidelberg, Berlin, Heidelberg, 1989.

[74] B. L. Golden, E. A. Wasil, and P. T. Harker. The Analytic Hierarchy Process: Applications and Studies. Springer Berlin Heidelberg, Berlin, Heidelberg, 1989.

[75] P. Grošelj, L. Zadnik Stirn, N. Ayrilmis, and M. K. Kuzman. Comparison of some aggregation techniques using group analytic hierarchy process. *Expert Systems with Applications*, 42(4):2198–2204, March 2015.

[76] P. T. Harker. Alternative modes of questioning in the analytic hierarchy process. *Mathematical Modelling*, 9(3):353 – 360, 1987.

[77] P. T Harker and L Vargas. The theory of ratio scale estimation: Saaty's analytic hierarchy process. *Management Science*, 33(11):1383–1403, November 1987.

[78] E. A. Hefnawy and A. S. Mohammed. Review of different methods for deriving weights in The Analytic Hierarchy Process. *International Journal of Analytic Hierarchy Process*, 6(1):92 – 123, 2014.

[79] W. Ho and X. Ma. The state-of-the-art integrations and applications of the analytic hierarchy process. *European Journal of Operational Research*, 267:399–414, 2018.

[80] R. D. Holder. Some Comments on the Analytic Hierarchy Process. *The Journal of the Operational Research Society*, 41(11):1073–1076, November 1990.

[81] S. S. Hosseinian, H. Navidi, and A. Hajfathaliha. A New Linear Programming Method for Weights Generation and Group Decision Making in the Analytic Hierarchy Process. *Group Decision and Negotiation*, 21(3):233–254, November 2009.

[82] N. V. Hovanov, J. W. Kolari, and M. V. Sokolov. Deriving weights from general pairwise comparison matrices. *Mathematical Social Sciences*, 55(2):205 – 220, 2008.

[83] Y. S. Huang, J. T. Liao, and Z. L. Lin. A study on aggregation of group decisions. *Systems Research and Behavioral Science*, 26(4):445–454, July 2009.

[84] E. K. R. E. Huizingh and H. C. J. Vrolijk. A Comparison of Verbal and Numerical Judgments in the Analytic Hierarchy Process. *Organizational Behavior and Human Decision Processes*, 70(3):237–247, June 1997.

[85] Y. Iida. Ordinality consistency test about items and notation of a pairwise comparison matrix in AHP. In *Proceedings of the International Symposium on the Analytic Hierarchy Process*, 2009.

[86] Y. Iida. The number of circular triads in a pairwise comparison matrix and a consistency test in the AHP. *Journal of the Operations Research Society of Japan*, 52(2):174–185, 2009.

[87] A. Ishizaka, D. Balkenborg, and T. Kaplan. Influence of aggregation and measurement scale on ranking a compromise alternative in AHP. *Journal of the Operational Research Society*, 62(4):700–710, March 2010.

[88] A. Ishizaka and A. Labib. Review of the main developments in the Analytic Hierarchy Process. *Expert Systems with Applications*, 38(11):14336–14345, October 2011.

[89] P. H. Iz and L. R. Gardiner. Analysis of multiple criteria decision support systems for cooperative groups. *Group Decision and Negotiation*, 2(1):61–79, Mar 1993.

[90] D. Jato-Espino, E. Castillo-Lopez, J. Rodriguez-Hernandez, and J. C. Canteras-Jordana. A review of application of multi-criteria decision making methods in construction. *Automation in construction*, 45:151–162, September 2014.

[91] R. E. Jensen. Comparisons of consensus methods for priority ranking problems. *Decision Sciences*, 17(2):195–211, 1986.

[92] R. E. Jensen and T. E. Hicks. Ordinal data AHP analysis: A proposed coefficient of consistency and a nonparametric test. *Math. Comput. Model.*, 17(4-5):135–150, February 1993.

[93] C. Kahraman, editor. *Fuzzy Multi-Criteria Decision Making: Theory and Applications with Recent Developments (Springer Optimization and Its Applications, 16)*. Springer, July 2008.

[94] P. Kazibudzki. Redefinition of triad's inconsistency and its impact on the consistency measurement of pairwise comparison matrix. 2016:71–78, 03 2016.

[95] M.G. Kendall and B. Smith. On the method of paired comparisons. *Biometrika*, 31(3/4):324–345, 1940.

[96] G. Kéri. On qualitatively consistent, transitive and contradictory judgment matrices emerging from multiattribute decision procedures. *Central European Journal of Operations Research*, 19(2):215–224, February 2010.

[97] G. Khatwani and A. K. Kar. Improving the Cosine Consistency Index for the analytic hierarchy process for solving multi-criteria decision making problems. *Applied Computing and Informatics*, 13(2):118–129, July 2017.

[98] W. W. Koczkodaj. A new definition of consistency of pairwise comparisons. *Math. Comput. Model.*, 18(7):79–84, October 1993.

[99] W. W. Koczkodaj. *Pairwise Comparisons Rating Scale Paradox*, pages 1–9. Springer Berlin Heidelberg, Berlin, Heidelberg, 2016.

[100] W. W. Koczkodaj, M. W. Herman, and M. Orlowski. Managing Null Entries in Pairwise Comparisons. *Knowledge and Information Systems*, 1(1):119–125, 1999.

[101] W. W. Koczkodaj, K. Kułakowski, and A. Ligęza. On the quality evaluation of scientific entities in Poland supported by consistency-driven pairwise comparisons method. *Scientometrics*, 99(3):911–926, 2014.

[102] W. W. Koczkodaj and M. Orłowski. Computing a consistent approximation to a generalized pairwise comparisons matrix. *Computers & Mathematics with Applications*, 37(3):79–85, 1999.

[103] W. W. Koczkodaj and S. J. Szarek. On distance-based inconsistency reduction algorithms for pairwise comparisons. *Logic Journal of the IGPL*, 18(6):859–869, October 2010.

[104] W. W. Koczkodaj and R. Szwarc. On axiomatization of inconsistency indicators for pairwise comparisons. *Fundamenta Informaticae*, 4(132):485–500, 2014.

[105] W. W. Koczkodaj and J. Szybowski. Pairwise comparisons simplified. *Applied Mathematics and Computation*, 253:387 – 394, 2015.

[106] W. W. Koczkodaj and R. Urban. Axiomatization of inconsistency indicators for pairwise comparisons. *International Journal of Approximate Reasoning*, 94:18–29, March 2018.

[107] G. Kou and C. Lin. A cosine maximization method for the priority vector derivation in AHP. *European Journal of Operational Research*, 235(1):225–232, May 2014.

[108] K. Kułakowski. Heuristic Rating Estimation Approach to The Pairwise Comparisons Method. *Fundamenta Informaticae*, 133:367–386, 2014.

[109] K. Kułakowski. Notes on Order Preservation and Consistency in AHP. *European Journal of Operational Research*, 245(1):333–337, 2015.

[110] K. Kułakowski. On the properties of the priority deriving procedure in the pairwise comparisons method. *Fundamenta Informaticae*, 139(4):403 – 419, July 2015.

[111] K. Kułakowski. Notes on the existence of a solution in the pairwise comparisons method using the heuristic rating estimation approach. *Annals of Mathematics and Artificial Intelligence*, 77(1):105–121, 2016.

[112] K. Kułakowski. Notes on the existence of a solution in the pairwise comparisons method using the heuristic rating estimation approach. *Annals of Mathematics and Artificial Intelligence*, 77(1):105–121, 2016.

[113] K. Kułakowski. Inconsistency in the ordinal pairwise comparisons method with and without ties. *European Journal of Operational Research*, 270(1):314 – 327, 2018.

[114] K. Kulakowski. On the geometric mean method for incomplete pairwise comparisons (preprint arxiv.org). *CoRR*, abs/1905.04609, 2019.

[115] K. Kułakowski, K. Grobler-Dębska, and J. Wąs. Heuristic rating estimation: geometric approach. *Journal of Global Optimization*, 62, 2014.

[116] K. Kułakowski and A. Kedzior. Some Remarks on the Mean-Based Prioritization Methods in AHP. In Ngoc-Thanh Nguyen, Lazaros Iliadis, Yannis Manolopoulos, and Bogdan Trawiński, editors, *Lecture Notes In Computer Science, Computational Collective Intelligence: 8th International Conference, ICCCI 2016, Halkidiki, Greece, September 28-30, 2016. Proceedings, Part I*, pages 434–443. Springer International Publishing, 2016.

[117] K. Kułakowski, J. Mazurek, J. Ramík, and M. Soltys. When is the condition of order preservation met? *European Journal of Operational Research*, 277(1):248–254, August 2019.

[118] K. Kułakowski and J. Szybowski. The new triad based inconsistency indices for pairwise comparisons. *Procedia Computer Science*, 35(0):1132 – 1137, 2014.

[119] K. Kułakowski and D. Talaga. Inconsistency indices for incomplete pairwise comparisons matrices. *International Journal of General Systems*, 0(0):1–27, 2020.

[120] R. J. Kuo, L. Y. Lee, and Tung-Lai Hu. Developing a supplier selection system through integrating fuzzy ahp and fuzzy dea: a case study on an auto lighting system company in taiwan. *Production Planning & Control*, 21(5):468–484, 2010.

[121] M. Kwiesielewicz. The logarithmic least squares and the generalized pseudoinverse in estimating ratios. *European Journal of Operational Research*, 93(3):611–619, September 1996.

[122] M. Kwiesielewicz. A note on the fuzzy extension of Saaty's priority theory. *Fuzzy Sets and Systems*, 95:161–172, 1998.

[123] Kwang H. Lee. *First Course on Fuzzy Theory and Applications*. Springer Berlin Heidelberg, Berlin, Heidelberg, 2005.

[124] F. Liang, M. Brunelli, and J. Rezaei. Consistency issues in the best worst method: Measurements and thresholds. *Omega*, page 102175, December 2019.

[125] F. A. Lootsma. Conflict resolution via pairwise comparison of concessions. *European Journal of Operational Research*, 40(1):109–116, May 1989.

[126] M. Lundy, S. Siraj, and S. Greco. The mathematical equivalence of the "spanning tree" and row geometric mean preference vectors and its implications for preference analysis. *European Journal of Operational Research*, (257):197–208, September 2017.

[127] D. Ma and X. Zheng. 9/9-9/1 scale method of ahp. In *In 2nd Int. Symposium on AHP pp. 197–202.*, volume 1. University of Pittsburgh, University of Pittsburgh, 1991.

[128] R. Merris. Laplacian matrices of graphs: a survey. *Linear Algebra and its Applications*, 197-198:143–176, January 1994.

[129] C. Meyer. *Matrix Analysis and Applied Linear Algebra*. SIAM: Society for Industrial and Applied Mathematics, April 2000.

[130] L. Mikhailov. Deriving priorities from fuzzy pairwise comparison judgements. *Fuzzy Sets and Systems*, 134(3):365–385, March 2003.

[131] L. Mikhailov. Group prioritization in the AHP by fuzzy preference programming method. *Computers & operations research*, 31(2):293–301, 2004.

[132] G. A. Miller. The magical number seven, plus or minus two: some limits on our capacity for processing information. *Psychological Review*, 63(2):81–97, 1956.

[133] B. Mohar. Laplace eigenvalues of graphs—a survey. *Discrete Mathematics*, 109(1-3):171–183, November 1992.

[134] J. M. Moreno-Jiménez, J. A. Joven, A. R. Pirla, and A. T. Lanuza. A Spreadsheet Module for Consistent Consensus Building in AHP-Group Decision Making. *Group Decision and Negotiation*, 14(2):89–108, Mar 2005.

[135] J. V. Neumann and O. Morgenstern. *Theory of Games and Economic Behavior*. Princeton University Press, 1947.

[136] T. Nguyen. Methods in Ranking Fuzzy Numbers: A Unified Index and Comparative Reviews. *Complexity*, 2017(4):1–13, 2017.

[137] G. Oliva, R. Setola, and A. Scala. Sparse and distributed Analytic Hierarchy Process. *Automatica*, 85:211–220, November 2017.

[138] W. Ossadnik, S. Schinke, and R. H. Kaspar. Group Aggregation Techniques for Analytic Hierarchy Process and Analytic Network Process: A Comparative Analysis. *Group Decision and Negotiation*, 25(2):421–457, July 2015.

[139] A. N. Patil, N.G.P. Bhale, N. Raikar, and M. Prabhakaran. Car Selection Using Hybrid Fuzzy AHP and Grey Relation Analysis Approach. *International Journal of Perofrmability Engineering*, 13(5):569–576, 2017.

[140] J. I. Peláez and M. T. Lamata. A new measure of consistency for positive reciprocal matrices. *Computers & Mathematics with Applications*, 46(12):1839 – 1845, 2003.

[141] S. Pemmaraju and S. Skiena. *Computational Discrete Mathematics - Combinatorics and Graph Theory with Mathematica*. Cambridge University Press, January 2003.

[142] C A. Pérez and V. R. Montequín. Integrating Analytic Hierarchy Process (AHP) and Balanced Scorecard (BSC) Framework for Sustainable Business in a Software Factory in the Financial Sector. *Sustainability*, 9(4):486, 2017.

[143] R. J. Plemmons. M-matrix characterizations. I - nonsingular M-matrices. *Linear Algebra and its Applications*, 18(2):175–188, December 1976.

[144] Plutarch. *The complete works of Plutarch: essays and miscelanies*, volume III. New York: Crowel, 1909.

[145] E. E. Poor, C. Loucks, A. Jakes, and D. L. Urban. Comparing Habitat Suitability and Connectivity Modeling Methods for Conserving Pronghorn Migrations. *PLOS ONE*, 7(11):e49390, November 2012.

[146] A. Quarteroni, R. Sacco, and F. Saleri. *Numerical mathematics*. Springer Verlag, 2000.

[147] J. Ramik. A decision system using ANP and fuzzy inputs. *International Journal of Innovative Computing*, 3(4):825–837, Aug 2007.

[148] J. Ramik and P. Korviny. Inconsistency of pair-wise comparison matrix with fuzzy elements based on geometric mean. *Elsevier*, 161:1604–1613, 2010.

[149] J. Ramík and M. Vlach. Measuring consistency and inconsistency of pair comparison systems. *Kybernetika*, 49(3):465–486, 2013.

[150] H. M. Regan, M. Colyvan, and L. Markovchick-Nicholls. A formal model for consensus and negotiation in environmental management. *Journal of environmental management*, 80(2):167–176, July 2006.

[151] J. Rezaei. Best-worst multi-criteria decision-making method. *Omega*, 53(C):49–57, June 2015.

[152] F. S. Roberts. *Measurement Theory with Applications to Decisionmaking, Utility, and the Social Sciences*. Encyclopedia of mathematics and its applications. Cambridge University Press, 1985.

[153] L. Rutkowski. *Computational intelligence : methods and techniques*. Berlin : Springer, Berlin, Heidelberg, 2008.

[154] D. G. Saari and V. R. Merlin. The Copeland method. *Economic Theory*, 8(1):51–76, February 1996.

[155] T. L. Saaty. A scaling method for priorities in hierarchical structures. *Journal of Mathematical Psychology*, 15(3):234 – 281, 1977.

[156] T. L. Saaty. Modeling unstructured decision problems — the theory of analytical hierarchies. *Mathematics and Computers in Simulation*, 20(3):147–158, 1978.

[157] T. L. Saaty. *The analytic hierarchy process: planning, priority setting, resource allocation*. McGrawHill International Book Co., New York; London, 1980.

[158] T. L. Saaty. *Decision Making for Leaders*. Lifetime Learning Publications divisions, Wadsworth, Belmont, CA, 1982.

[159] T. L. Saaty. Axiomatic Foundation of the Analytic Hierarchy Process. *Management Science*, 32(7):841–855, 1986.

[160] T. L. Saaty. *Group Decision Making and the AHP*, pages 59–67. Springer Berlin Heidelberg, Berlin, Heidelberg, 1989.

[161] T. L. Saaty. An Exposition on the AHP in Reply to the Paper "Remarks on the Analytic Hierarchy Process". *Management Science*, 36(3):259–268, March 1990.

[162] T. L. Saaty. The analytic hierarchy and analytic network processes for the measurement of intangible criteria and for decision-making. In *Multiple Criteria Decision Analysis: State of the Art Surveys*, volume 78 of *International Series in Operations Research and Management Science*, pages 345–405. Springer New York, 2005.

[163] T. L. Saaty. Relative Measurement and Its Generalization in Decision Making. Why Pairwise Comparisons are Central in Mathematics for the Measurement of Intangible Factors. The Analytic Hierarchy/Network Process. *Estadística e Investigación Operativa / Statistics and Operations Research (RACSAM)*, 102:251–318, November 2008.

[164] T. L. Saaty. *Theory and Applications of the Analytic Network Process*. RWS Publications, 2009.

[165] T. L. Saaty. On the Measurement of Intangibles. A Principal Eigenvector Approach to Relative Measurement Derived from Paired Comparisons. *Notices of the American Mathematical Society*, 60(02):192–208, February 2013.

[166] T. L. Saaty. *The Neural Network Processor (NNP) - The Fundamental Equation of Neural Response and Consciousness*. RWS Publications, 2015.

[167] T. L. Saaty and K. Kułakowski. Axioms of the analytic hierarchy process (AHP) and its generalization to dependence and feedback: The analytic network process (ANP). *CoRR*, abs/1605.05777, 2016.

[168] T. L. Saaty and M. S. Ozdemir. Why the magic number seven plus or minus two. *Math. Comput. Model.*, 38(3-4):233–244, 2003.

[169] T. L. Saaty and L. G. Vargas. *Models, Methods, Concepts & Applications of the Analytic Hierarchy Process*, volume 175 of *International Series in Operations Research & Management Science*. Springer US, Boston, MA, 2012.

[170] T. L. Saaty and L. G. Vargas. *Decision Making with the Analytic Network Process*, volume 195 of *Economic, Political, Social and Technological Applications with Benefits, Opportunities, Costs and Risks*. Springer Science and Business Media, Boston, MA, May 2013.

[171] T. L. Saaty, L. G. Vargas, and H. J. Zoffer. A structured scientific solution to the Israeli–Palestinian conflict: the analytic hierarchy process approach. *Decision Analytics*, 2(1):7, Sep 2015.

[172] A. A. Salo and R. P. Hämäläinen. Preference programming through approximate ratio comparisons. *European Journal of Operational Research*, 82(3):458 – 475, 1995.

[173] A. A. Salo and R. P. Hämäläinen. On the measurement of preferences in the analytic hierarchy process. *Journal of Multi-Criteria Decision Analysis*, 6(6):309–319, 1997.

[174] K. Sangheetha Sri Prakrash, H. Hassena Begum, and M. Pavithra Anand Babu. Ranking Of Triangular Fuzzy Number Method To Solve An Unbalanced Assignment Problem. *Journal of Global Research in Mathematical Archives*, 2(8):06–11, 2014.

[175] B. Schoner and W. C. Wedley. Alternative Scales in AHP. In *Improving Decision Making in Organisations*, pages 345–354. Springer, Berlin, Heidelberg, Berlin, Heidelberg, 1989.

[176] S. Shiraishi and T. Obata. On a MaximizationProblem Arising from a Positive Reciprocal Matrix in AHP. *Bulletin of Informatics and Cybernetics*, 34(2):91–96, December 2002.

[177] S. Shiraishi, T. Obata, and M. Daigo. Properties of a positive reciprocal matrix and their application to AHP. *Journal of the Operations Research Society of Japan*, 41(3):404–414, September 1998.

[178] S. Siraj, L. Mikhailov, and J. Keane. A heuristic method to rectify intransitive judgments in pairwise comparison matrices. *European Journal of Operational Research*, 216(2):420–428, January 2012.

[179] S. Siraj, L. Mikhailov, and J. A. Keane. Enumerating all spanning trees for pairwise comparisons. *Computers and Operations Research*, 39(2):191–199, February 2012.

[180] S. Siraj, L. Mikhailov, and J. A. Keane. Preference elicitation from inconsistent judgments using multi-objective optimization. *European Journal of Operational Research*, 220(2):461–471, July 2012.

[181] W. E. Stein and P. J. Mizzi. The harmonic consistency index for the Analytic Hierarchy Process. *European Journal of Operational Research*, 177(1):488–497, February 2007.

[182] J. Stoklasa, V. Jandová, and J. Talasová. Weak consistency in Saaty's AHP - evaluating creative work outcomes of Czech Art Colleges. *Neural Network World*, 1(13):61–77, 2013.

[183] J. Stoklasa and T. Talášek of. AHP and weak consistency in the evaluation of works of art - a case study of a large problem. *International Journal of Business Innovation and Research*, 11(1):60, 2016.

[184] R. Storn and K. Price. Differential Evolution – A Simple and Efficient Heuristic for global Optimization over Continuous Spaces. *Journal of Global Optimization*, 11(4):341–359, 1997.

[185] N. Subramanian and R. Ramanathan. A review of applications of Analytic Hierarchy Process in operations management. *International Journal of Production Economics*, 138(2):215–241, August 2012.

[186] L. Sun and B. S. Greenberg. Multicriteria Group Decision Making: Optimal Priority Synthesis from Pairwise Comparisons. *Journal of Optimization Theory and Applications*, 130(2):317–339, Aug 2006.

[187] K. Suzumura, K. J. Arrow, and A. K. Sen. *Handbook of Social Choice & Welfare*. Elsevier Science Inc., 2010.

[188] J. Szybowski. The Cycle Inconsistency Index in Pairwise Comparisons Matrices. *Procedia Computer Science*, 96:879–886, January 2016.

[189] E. Takeda and P. Yu. Assessing priority weights from subsets of pairwise comparisons in multiple criteria optimization problems. *European Journal of Operational Research*, 86(2):315–331, October 1995.

[190] J. A. Thorpe. *Elementary Topics in Differential Geometry*. Undergraduate Texts in Mathematics. Springer Science & Business Media, New York, NY, 1979.

[191] K. Tone. Logarithmic Least Squares Method for Incomplete Pairwise Comparisons in the Analytic Hierarchy Process. Technical Report 94-B-2, Saitama University, Institute for Policy Science Research, Urawa, Saitama, 338, Japan, December 1993.

[192] E. Triantaphyllou and S. H. Mann. An examination of the effectiveness of multi-dimensional decision-making methods - A decision-making paradox. *Decision Support Systems*, 5(3):303–312, 1989.

[193] V. Tsyganok. Investigation of the aggregation effectiveness of expert estimates obtained by the pairwise comparison method. *Mathematical and Computer Modelling*, 52(3-4):538–544, August 2010.

[194] P. J. M. Van Laarhoven and W. Pedrycz. A fuzzy extension of Saaty's priority theory. *Fuzzy Sets and Systems*, 11(1-3):229–241, 1983.

[195] Y. Wang. Ranking triangle and trapezoidal fuzzy numbers based on the relative preference relation. *Applied Mathematical Modelling*, 39(2):586 – 599, 2015.

[196] Y. Wind and T. L. Saaty. Marketing Applications of the Analytic Hierarchy Process. *Management Science*, 26(7):641–658, 1980.

[197] K. K. F. Yuen. Membership Maximization Prioritization Methods for Fuzzy Analytic Hierarchy Process. *Fuzzy Optimization and Decision Making*, 11(2):113–133, June 2012.

[198] K. K. F. Yuen. The Primitive Cognitive Network Process: Comparisons With The Analytic Hierarchy Process. *International Journal of Information Technology & Decision Making*, 10(04):659–680, April 2012.

[199] K. K. F. Yuen. Fuzzy Cognitive Network Process: Comparisons With Fuzzy Analytic Hierarchy Process in New Product Development Strategy. *IEEE Transactions on Fuzzy Systems*, 22(3):597–610, 2014.

[200] K. K. F. Yuen. The primitive cognitive network process in healthcare and medical decision making: Comparisons with the analytic hierarchy process. *Applied Soft Computing*, 14, Part A(0):109 – 119, 2014.

[201] F. Zahedi. A simulation study of estimation methods in the analytic hierarchy process. *Elsevier*, 20(6):347–354, January 1986.

[202] S. Zhou and P. Yang. Risk management in distributed wind energy implementing analytic hierarchy process. *Renewable Energy*, 150:616–623, 2020. cited By 0.

[203] K. Zhu, Y. Jing, and D. Chang. A discussion on Extent Analysis Method and applications of fuzzy AHP. *European Journal of Operational Research*, 116(2):450 – 456, 1999.

Index

Printed and bound by CPI Group (UK) Ltd, Croydon, CR0 4YY

24/10/2024

01778303-0001